Quarterly Essay

CONTENTS

Quarterly Essay is published four times a year by Black Inc., an imprint of Schwartz Media Pty Ltd. Publisher: Morry Schwartz.

ISBN 978-1-86395-560-7 ISSN 1832-0953

Subscriptions – 1 year (4 issues): $49 within Australia incl. GST. Outside Australia $79.
2 years (8 issues): $95 within Australia incl. GST. Outside Australia $155.

Payment may be made by Mastercard or Visa, or by cheque made out to Schwartz Media. Payment includes postage and handling.

To subscribe, fill out and post the subscription card or form inside this issue, or subscribe online:

www.quarterlyessay.com
subscribe@blackincbooks.com
Phone: 61 3 9486 0288

Correspondence should be addressed to:

The Editor, Quarterly Essay
37–39 Langridge Street
Collingwood VIC 3066 Australia
Phone: 61 3 9486 0288 / Fax: 61 3 9486 0244
Email: quarterlyessay@blackincbooks.com

Editor: Chris Feik. Management: Sophy Williams, Jess Tran. Publicity: Elisabeth Young. Design: Guy Mirabella. Assistant Editor/Production Coordinator: Nikola Lusk. Typesetting: Duncan Blachford.

Printed by Griffin Press, Australia. The paper used to produce this book comes from wood grown in sustainable forests.

US AND THEM

On the Importance of Animals

Anna Krien

It was around midnight when I got off the tram at the last stop in Melbourne's north. As the doors locked behind me, two men, one bare-chested, ran across the street, traffic swerving, and stood in front of the tram, arms crossed. Their eyes were opaque, faces shiny with sweat. "Take us to the city," one yelled, while the other went around to the driver's window and started banging on it with his fist.

"This is the last stop," the driver pleaded through his microphone. "Please, this is the last stop. Let me finish."

The shirtless man reared up, kicking the tram. "I said, take us to the city!"

In the shadows, I weighed it up. The driver was safe inside the tram; he could phone for help. I, however, was on the outside with them. If I called them off, they'd start in on me.

I slipped away, leaving the yelling men and the tinny "Please move away from the tram" behind me. This is not who I wanted to become, someone who measures the weight of an injustice before intervening, but cities can do that to a person. Walking beside the train tracks to my home, I was unusually nervous. A couple of months earlier, a girl's throat was slit near here, and last week a pack of guys attacked three girls on a street corner close by. Then, on the footpath ahead of me, a large black dog appeared, bright red collar around its neck. No owner to be seen. It must have slipped out of its backyard for a midnight sniff around the block. As I approached, we cocked our heads at one another and then the dog picked up my pace and trotted alongside. Together we walked, its nails clopping on the cement. When we got to my gate, I unlatched it and closed it behind me, leaving the dog on the footpath. I dropped my hand over the fence, scratching its head, and for a minute we stood like that. Then, with a nod, we went our separate ways.

It was the smallest thing. And I know it could have been the other way around: I could have come across a snarling dog, or those two men at the tram could have had an equally vicious hound with them, and then a nice young man could have appeared out of nowhere and walked me home. But that's not what happened. I met an amicable creature in a hostile night and that creature was a dog.

A few years ago a letter was read out on America's National Public Radio about a stray orange cat that wandered into a Michigan prison where many of the male prisoners were serving life sentences. Troy Chapman, the inmate who wrote the letter, said that when he knelt down to pat the cat, it was the first time in twenty years that he had touched an animal. He described how, over the next few weeks, the cat broke down the tough prison culture. Men who had never exchanged words chatted while they petted the cat; one groomed the animal and pulled burrs from its matted coat; others smuggled saucers of milk and scraps of meat from the dining hall and prison kitchen to it, carefully placing the scraps under the dumpster so the seagulls couldn't reach them.

"There's a lot of talk about what's wrong with prisons in America," Chapman said in his letter. "We need more programs; we need more psychologists or treatment of various kinds. Some even talk about making prisons more kind, but I think what we really need is a chance to practise kindness ourselves. Not receive it, but give it."

I read this letter, written by a man incarcerated for life after being convicted of second-degree murder, many times while writing this essay. I read it when my confidence wavered, when I wondered about the seriousness of my arguments. This happened quite a lot. I mean, really, a *Quarterly Essay* about *animals*? In a series of intelligent analysis that has looked at the looming superpower that is China, the origins of conservatism, the history wars, indigenous Australians, climate change, American and local politics – here I am, writing about animals?

Last year, in Fairfax's *Good Weekend*, the novelist Charlotte Wood wrote of her discomfort with people who gushed over animals and a society that humanises them beyond recognition. "Our culture is drenched in anthropomorphic slush," Wood wrote:

> But I find most of it troubling because it seems so disrespectful. Denying the creature's essential nature – its very animality – is surely an act not of admiration, but subjugation. To downplay the differences between species is to promote the assumption that "humans will only accept what is like themselves," as American scholar Shelly R. Scott puts it.

The more we sentimentalise animals on the one hand, Wood suggested, the more we brutalise them on the other.

Everywhere you look in our society, there are animals. Well, not *actual* animals, but caricatures of animals selling us mobile-phone plans, toilet paper, cars, interest rates. As projections, animals proliferate in the flat-screened zoos of cinema, television, YouTube and screen savers, trapped in Pixar worlds, serving as light relief at the end of the day. In shops, you can buy animals cast in clay, plastic, porcelain, and you may hold your new purchase close, cradle it on the way home and place it carefully on the bookshelf, a tiny guardian to watch over your world. Hell, animals are even in our dreams – we wake up talking of lions, snakes and wolves, wondering what they could possibly mean.

2011 was a big year for animals. Asian manufacturers flooded the

Western fashion market with leopard-print garments; hipster bands named themselves after birds, deer and horses; and in America the Wrangler Jeans' "We Are Animals" campaign featured moody photos of muddied models, wearing denim jeans, strewn like seaweed across boulders, creeks and roads.

Locally, Kevin Rudd underwent intensive heart surgery. "Time for a bit of a grease and oil change myself; I'll be having aortic valve replacement surgery," said Rudd to reporters. "The docs have said, 'Kev, it's time for a new one.' And so, being the obedient soul that I am, I've decided to take their advice." Rudd emerged with a cow valve inserted inside his chest.

At the MCG, the Australian Football League trialled the use of two wedge-tailed eagles to deter seagulls from hobnobbing on the brightly lit oval during a semi-final game.

In a ceremony on the lawn at the Australian War Memorial in Canberra, "Sarbi," an explosive-detection dog working with the defence force in Afghanistan, was awarded a Purple Cross bravery medal. "For the courage she has shown while serving her country during her time in Afghanistan," said the president of the RSPCA, Lynne Bradshaw, solemnly hanging the medal around the black labrador cross's neck.

Around the world, footage of laboratory chimpanzees setting foot outside on grass for the first time became an internet sensation. For over thirty years, the chimps had been kept in laboratories and injected with diseases such as HIV and hepatitis. The images of the primates clustering around the open doorway of their new sanctuary, hugging (yes, *hugging*) and then running, whooping, across the lawn could unnerve even the hardest of pro–animal testing hearts.

Then, one evening in Melbourne's western suburbs, a pit bull mastiff cross ran up the driveway of a St Albans house, chasing the occupants inside and following them into the lounge room, where it mauled four-year-old Ayen Chol to death. "These types of dogs have lost their right to exist in Victoria," said the state's agriculture and food security minister, Peter Walsh, after the tragedy.

And, finally, the cows. In May, in its program "Bloody Business," ABC's *Four Corners* aired footage of Australian cows being grossly mistreated in eleven Indonesian abattoirs. The wave of revulsion that followed – the inundation of the news media, the overloading of the RSPCA and Animals Australia websites within hours of the report, the thousands of protesters taking to the streets, MPs calling for a conscience vote or threatening to revolt against the party line, and airwaves blocked with people calling for a ban on the trade – forced a ban on the live export of cattle to Indonesia. The veteran political writer Michelle Grattan described parliamentary debate after the screening as the most passionate she had seen in years.

The public response seemed a clear assertion that what had happened was wrong and intolerable. But the story of the live cattle trade is more complex than it seems. And so too is our nuanced and often contradictory relationship with animals. It seems most of us have a minor clause inside us on the treatment of animals – a "that's not allowed" but "that's okay." We have our limits and our permissions. But the categories are becoming more and more blurred. After all, how can we allow one act and not another?

For example, the cattle slaughtered in Indonesia are, to all intents and purposes, objects. "Things" that suddenly became subjects in the glare of a video camera. And to what end? For killing standards to be raised to Western-approved levels so that cattle can safely become objects once more?

Let's just say this recent controversy succeeds in raising the bar, that slaughtering in Indonesia and the other thirty-odd countries to which Australia exports live animals, either by ship or plane, becomes "world standard." Death by stunning is implemented across the board. There is Meat and Livestock Australia's "in the ute, not the boot" campaign, launched in Egypt after Animals Australia (yes, the animal welfare group does seem to be the regulator here) brought back footage of men stuffing sheep into the boot of their car or tying them on the roof rack; the many temporary export bans; the thirty-odd years of intermittent political scrambling followed by political and public amnesia; the Memorandums of Understanding and minor adjustments to the trade – let's say all this sinks in.

Animals around the world are not kicked or gouged, cows do not – as one Indonesian abattoir was found guilty of five years ago – have hoses shoved in their mouths and anuses before slaughter, filling them with water so the meat will be heavier and sell for more at the market. Instead they're calmly led to the killing floor, stunned and killed on the first cut. Will that quell the growing unease among many Westerners, a nagging sense, as the Nobel laureate J.M. Coetzee wrote in the *Sydney Morning Herald* late last year, that "Something has gone badly wrong in relations between human beings and other animals"?

> Even people who take their lead from Genesis, from its assurance that God has granted us dominion over the beasts in order to feed ourselves, suffer nagging doubts whether factory farming and a food industry operating on an industrial scale to turn living animals into what are euphemistically called "animal products" are quite what God had in mind.

"In the eyes of a butcher a horse is already dead," wrote Georges Bataille. When I read this, I was reminded of the opaque eyes of the men who accosted the tram driver. Cold slippery eyes that didn't see people, eyes that, for whatever reason, held no warmth and saw the world, the rest of us, as being at their service. What happens when we turn this look, this strangely blind and indifferent gaze, into an industry?

How can we maintain that animals are things but also beings?

By law in Australia, as in much of the world, animals are "things" and yet they are things with welfare legislation – although some animals enjoy more comfort than others. Livestock, for example, numbering around 500 million in Australia, do not enjoy the same level of protection as companion animals – they are often exempt from rulings on access to sufficient exercise and natural light.

In 1992, Switzerland amended its constitution to recognise animals as "beings" rather than "things." The amendment, however, was short-lived after the entire constitution was rewritten to leave animals in a kind of status limbo. And two years ago in Spain, higher primates such as apes, chimpanzees, bonobos and orangutans were granted certain "rights" by the Spanish parliament, the first national legislature to do so.

"This is either a ridiculous society or a dislocated one," the archbishop of Pamplona told *Time* magazine in 2008, reiterating the Catholic Church's argument that the move to grant apes fundamental rights eroded the Biblical hierarchy which gives humans dominion over the earth. "Asking for human rights for monkeys is like asking for bull rights for men."

Bull rights? Well, not yet, but last year Catalonia staged its last bullfight.

"The fact that some people are silly about animals cannot stop the topic being a serious one," wrote the philosopher Mary Midgely. Animals are like poetry — supple, simultaneously cryptic and revealing. They are like a patch of sunshine in a garden, on your bed, in a fading afternoon. And like poetry, there's also a whole lot of crap involved. For one, they're not poetry.

But we're drawn to them, and in many ways they're drawn to us — and this essay is a meditation on that meeting point. "As with most things that confuse me," wrote Charlotte Wood,

> I suspect simply paying closer attention might unlock the puzzling nature of human–animal bonds … for one species to trust another enough to lie down together and sleep — not because of their similarity but despite their difference — is, when you think about it, awe-inspiring. And perhaps it's in this quiet space that things might be revealed.

I once saw a fox wander down Swanston Street at 4 a.m., its red flame of a tail hovering like a healing hand over the city gutter. And yes, I know it is the wrong type of animal. A pest. But something in me swelled for it.

Another time I watched a tough and tattooed security guard at a XXX club feed the pigeons, those "rats of the sky." They were cooing around his sneakers, a couple even flying up to his outstretched hands, perching on his fingers. It was as if his face had cracked open to reveal an exquisite kindness, his immersion letting him forget the men down in the den behind him, slotting $2 coins to see naked girls behind a sheet of glass, a box of tissues beside them.

I'm afraid I never really grew up. I never got to that grown-up way of thinking that animals – be it in stories or in life – are either surrogates for humans or things to be used. And that is not to say that I don't eat them, don't wear them, don't ingest pills that have been tested on them. I have done all these things. I am not weighing up whether our treatment of animals is just, because it isn't. That age-old debate is a farce – deep down we all know it.

The real question is, just how much of this injustice are we prepared to live with? That is the premise of this essay. I am not going to be waylaid by the denigration of this issue that so often turns into a treatise on why humans are better than other animals – "It's our use of tools, our art and culture, our literary *prowess*, and therefore …" or why animals are like us – "Pigs have the intelligence of a three-year-old child, elephants mourn for their young, rats have empathy …"

Each of the following chapters deals with an encounter – from Indonesia to the laboratory to the Australian bush. In each encounter you will meet people distancing themselves from animals or intimating a false closeness in order to make use of them. Us and them; the lines that divide us are as mysterious as the lines that connect. But I will attempt to trace them nevertheless, and measure the weight of each injustice.

Here goes.

KILLING

Just days before my arrival in Jakarta, the streets of Indonesia ran with blood and stank of guts. DIY slaughters are a must for most Muslim festivals, in this case the Eid al-Adha. In the days leading up to it, goats were tethered in streets, some alone, others in temporary pens of "sacrificial animals" yet to be sold.

On the plane from Melbourne, I look through day-old Indonesian newspapers, at photos of the festival. Barefoot men stand in puddles of blood, while children helpfully stack the heads of goats. President Yudhoyono, I read, offered a 1.2-tonne cow for sacrifice at Jakarta's main mosque, while the local rock star "Ariel," in prison for appearing in and circulating celebrity sex tapes, donated a cow to the jail kitchen to share with his inmates.

At the airport, the Indonesian small talk I'd confidently recited for the past week evaporates as soon as I'm through the automatic doors. Suddenly mute, I let myself be hustled into a taxi and smother the familiar panic I get in foreign-speaking countries. Traffic on the highway into Jakarta is typical of most Southeast Asian highways; cars and motorbikes nudge, beep and weave.

Traffic here swarms, feeling its way forward.

Gazing at the motorbike riders, I remember a lecture by the autistic American cattle expert Temple Grandin. Talking to livestock farmers, she stated that cows perceive a man on horseback and a man on foot as two different things and that farmers need to habituate their livestock to both. I try to see the motorbike people as an animal might, turning them into modern mechanical versions of Pan, part bike, part man.

As my taxi siphons off into city streets, I see a man yank on a chain attached to a small monkey at the traffic lights. On the bricks behind them, a skinny orange cat jumps off the wall. The room I've booked is through a labyrinth of streets in a house with a red parrot and Alaskan huskies in cages. On my arrival, a local Indonesian woman is being baptised in the swimming pool.

*

"It's silly to say, but arriving in Indonesia is probably a culture shock for these cows," says Greg Pankhurst, who imports Australian cows to Indonesia. "Much like it is for us." I meet Greg in a hotel lobby in central Jakarta, a Westernised mecca of air-conditioning and shopping malls. Security is high. Taxi drivers pop their boots for inspection while guards run a length of pipe with a hook under the vehicle, checking for bombs. At the hotel, a doorman puts my bag on a wooden tray lined with blue linen and sends it through the X-ray machine in style.

Pankhurst describes the life of the purpose-bred Brahman cattle in Australia's Northern Territory. "In Australia, these cows have probably seen a person once or twice and they're mustered with a helicopter. They're wild. Then they arrive here, where there are people every-where! But," he adds, "they get used to it, like we do, I suppose. And they put on weight, which says they're not stressed."

Indonesia is now Australia's largest live export market for cows. Beginning in 1993, the trade grew to over 700,000 exported cattle in 2008. In recent years, however, the market has diminished, due to lim-itations put in place by the Indonesian government. Last year, Australian producers were caught off-guard when Indonesia announced that it would only accept cows 350 kilograms and below in a bid to create jobs in onshore feedlots. Shipments had to be reorganised as farmers scrambled to find new markets for their oversized cattle. Then, the Indonesian government capped the number of Australian import cattle at 500,000.

Recently, the trade has been capped again, this time to 280,000 cattle, shipped each year from Australia's northern ports to the main islands of Java and Sumatra on large wedding-cake tiered ships with names like the Ocean Drover.

From the Tanjung Priock port on the island of Java, the Panjang port in southernmost Sumatra and the Belawan port in northern Sumatra, the cattle are transported by trucks, trains, ferries and ships all over the vast archipelago to feedlots where they're fattened for at least sixty days on leftover corn, pineapple skins, tapioca, palm oil and other by-products bought from Indonesian farms.

Then, within twenty-odd hours, they are slaughtered, sold at wet markets (a fresh food market where you can buy chickens killed on the spot) and consumed – an efficient, neat and neighbourly trade, with one problem.

Cows are not radios. Nor are they tyres. Or television sets.

*

Within a week of Four Corners screening "Bloody Business," the federal agriculture minister, Joe Ludwig, buckled under public pressure and suspended the $300 million live cattle trade with Indonesia. At the Victorian Rural Press Club he told members he had made the decision after watching twenty hours of unedited footage provided by Animals Australia and Four Corners.

To begin with, I didn't watch it. I reasoned that I see enough horrible things and that I rarely eat meat. But the wave of revulsion, the strength of conviction, wore me down. So, wearing headphones, I watched the footage gathered by Animals Australia's key investigator, Lyn White, and the ABC journalists.

It's not something you can breathe through.

The method of casting – tying ropes around the cows' legs to trip them up – has them smashing their heads on slabs of concrete and metal bars, struggling to get up, rearing, straining against the ropes, sometimes managing to stand, only to slip-slip-slip on blood and shit, hooves clumsily ice-skating out from beneath them, legs getting caught in gutters and snapping, the guttural gurgling, the bizarre breaking of tails – grabbing, twisting them like ropes – sticking fingers in eyes, hoses up nostrils, and dancing – hands splayed, knives poised – around the blood-skating cows and slashing tendons to make them fall down, big confused animals, their eyes flaring, bellowing – the same sound you make when you scream into a pillow – a cow falling on a ramp and shutting down, legs awkward and jutting out, unable to comprehend what is expected of it, so another cow is urged over it, trampling it, and then the blunt knives hacking at throats, eyes blinking minutes after they should be dead, cows screaming and struggling to their feet, neck a gaping wound, bloody like a mouth, men walking away, not finishing the job, cows convulsing to death and finally the footage of a steer, tethered to a post, body shaking, as four of its herd are slit and skinned – *are they still alive?* – in front of it.

Here there were no "Judas goats," specially trained goats that once, in Australia, led cows and sheep quietly, submissively, to slaughter.

After seeing this I made the decision to visit Indonesia. At the same time I began to research the live cattle trade. I reasoned that this was one of those meeting points – where an animal found itself on the boundary between thing and being, where traditional and modern were colliding, and where the prospects for change in the treatment of animals could be examined close up.

*

"Hey, hello, miss honey, hey, how are you, miss honey?"

I'm at a bird market in central Jakarta, dressed modestly like all the guidebooks advise. A ring glints on my wedding finger. Above me, hundreds of rickety wooden cages filled with birds hang from hooks while other cages are stacked against the wall, but each time I stop to peer in at the birds, a cluster of men appears. So I whoosh past instead, past the chickens lying slumped on the wire floors, barely breathing, necks floppy like old rubber tubes. There's no water or feed in the cages. If they're bought, they'll be fed. But if no one wants to buy them, why let the feed go to waste? There are tiny cages packed with cockatiels, canaries, owls, macaws, rosellas, lovebirds, even common Indian mynah birds. Some cages are crowded with the smallest birds I have ever seen, the size of a thumbnail. There are chicks, dyed green, pink, red and blue. And the men, they keep following, "Where you going, miss honey?"

Much has been written about why women more than men are drawn to advocating for the welfare of animals. In 1990, the American author and animal activist Carol J. Adams's book *The Sexual Politics of Meat: A Feminist-Vegetarian Critical Theory* linked the treatment of animals as objects with patriarchal society's objectification of women, blacks and other minorities – and on the surface, this connection seems obvious. An ability to empathise with prey rather than predator, with the vulnerable rather than the perpetrator, with property rather than keeper, with being considered nothing more than a piece of meat, does seem more likely to be within the intuitive reach of females.

In the late 1800s, Frances Power Cobbe both founded the British Union for the Abolition of Vivisection and was a key member of the London National Society for Women's Suffrage, while Anna Kingsford became not only one of the first English women to obtain a degree in medicine, in 1880, but also the only student at the time to refuse to experiment on animals and still graduate.

Contemporary writers keen to make the link between women's rights and animal advocacy often refer to the philosopher Thomas Taylor's response to Mary Wollstonecraft's 1792 call for "the rights of woman." In reply, he wrote a mocking call for "the rights of brutes." To Taylor, granting rights to women was as absurd as, say, granting them to animals.

As I walk among the birds, overly aware of my gender and Western origins, making illogical turns to veer away from watching eyes, it's hard not to feel dramatic. I am free, am I not? Sure, I'm dressed modestly, but this is no burqa. My arms are not awkwardly bent like the wings around me, unable to stretch out in cane wicker cages. And yet I feel incredibly vulnerable. I fantasise about unlatching all the cages, of the birds tumbling out into the sky, of pressing a thumb on the necks of suffering, aching half-dead birds, snapping their spines, turning around, holding a man by the scruff, saying, "Stop leering, old man, or I'll do the same to you." But I keep walking. These birds will never leave their cages.

At the market's edge, one bird makes me stop.

In a cage hanging from a hook on the roof is a perfect bird. Its red, yellow and black feathers are blocks of colour, plumed in a geometric pattern across its trembling chest. The kind of creature that makes you wonder about the origin of things. A friend back home, a keen student of parrots, once said to me, "What's the point of a bird in a cage? It's like buying a painting, hanging it on the wall and throwing a sheet over it. The beauty of birds is their flight." And I agree with him, but how to stop all these eyes from wanting to own that picture of flight? How do I stop myself from wanting to own that picture of flight?

The bird hops, hypnotically. Kicking its feet forward, lunging its beak upwards, the perfect bird squawks at the dome of its cage. Over and over it does this, as though it is broken, stuck on a tune. "Hello, miss," a man breathes in the curve of my neck, a few men beside him looking up and down. "You like this bird?" I quickly step off the kerb into a tide of motorbikes. As I land, I find myself peering into a bin, at a pile of dead birds that didn't make it. The lucky ones?

*

"You take them one way and they want to go the other, you walk them over there and they just walk back," says a worker at the Dharma Jaya abattoir and feedlot in east Jakarta to my translator. He is talking about Australian cows. In spite of myself, I get a little flush of pride, of patriotism. As we speak, the lot is filling with cattle from the Northern Territory, big Brahmans coming off a truck, bellowing as they're nudged down the ramp to join others that have been fattening for three days now. These cows are standing and lying in ankle-deep mud and shit. The smell is so pungent and gag-inducing that I pinch my arm to distract my senses. Every now and then, we all stamp our feet, getting a brief rise out of the dozens of flies that have settled on us.

Dju (I have changed his name) has worked at this government-owned business for fifteen years. Like many Indonesians, he runs a little business on the side, and he gestures at three locally bred cows tied up, separate from the others in the feedlot. They look like calves next to the Australian cows, but they're fully grown males. Indonesians are not allowed to kill local female cows, which the government sees as crucial breeders in service of the aim to become self-sufficient in beef. But when demand for beef is high, supply is low and prices are good, as during the recent Ramadan, which saw a shortage of Australian cattle due to the ban, local farmers find it difficult to resist bringing their female cows in for slaughter.

Three months ago, representatives of the Australian cattle industry visited this abattoir and showed the workers some video footage. It was the same footage that had triggered the temporary Australian ban. I ask Dju what he thought of the treatment of the cows. "It's not right," he says, pausing. "But understandable." Australian cows, he says, are a handful. They won't be led, they're difficult to rope. Dju says his son is a slaughterman and is "exhausted" after a shift. "The Indonesian cow, you can lead him like a dog." At night, this abattoir slaughters between forty-five and one hundred head of cattle – the average nightly wage is 50,000 rupiah ($5) or a portion of beef. It operates almost every night of the year. The meat is then sold at wet markets in the mornings, where discerning buyers press their fingers into the meat, checking for warmth.

Dju takes us to the slaughtering area, which is beside the feedlot. The killing floor is a pockmarked cement slab lined with dirty tiled walls. Wide man-sized holes have been bashed through the walls to make openings. Chains and carcass hooks hang from the roof. Rose-tinted with last night's blood, the area has been hosed down. Printed on the wall in black painted letters is Arabic script. "Mecca faces that way," my translator explains. On the concrete slab, a man is crouched over a bloody puddle, rinsing out plastic drinking cups. "We sell everything," says Dju. "Even the faeces."

In the middle of the floor are two old Mark 1 "trip" boxes. In 2000, Meat and Livestock Australia (MLA) and Livecorp commissioned the design of the Mark 1 box after realising that the stress the Australian cattle underwent before slaughter was resulting in poor quality meat that was heavily discounted on the market. The metal box restrains the cow while workers rope its legs through a gap. Once the cow is roped, a worker opens the side of the box while others hold the ropes firm: the cow goes to take a step, the ropes yank and it trips, falling on its side on a sloped cement plinth, into a position of "lateral incumbency." Ideally the cow lands neatly in a ready-to-cut position, so that its throat can be slit in a single stroke. The Mark 1 boxes were installed throughout Indonesia (and in Brunei, Malaysia and the Middle East), followed by a series of locally made "copy" boxes, indicating the design's success, at least by local standards.

Most abattoirs in Indonesia use traditional slaughter methods – lassoing a cow to a metal ring on the floor with ten or so workers on the rope slowly dragging and pushing the animal onto the ground. This process can take up to thirty minutes with Australian cows. Until the ban, Australian cows were sent to around one hundred of these slaughterhouses, where killing varies from the traditional method to manually cutting the animals with the aid of the Mark 1 boxes. Only eight of the facilities had been fitted with stunners, the Australian standard for rendering an animal unconscious using an electrical current or a captive bolt pistol before throat-cutting. Of these, only a few used them regularly. The idea of putting in more stunners, according to Meat and Livestock Australia, was merely "aspirational." And even with a box, an ideal slaughter is not necessarily the norm.

As the footage on *Four Corners* revealed – zooming in on the industry signatures of MLA and Livecorp stencilled across the boxes – the cows sometimes tried to stand up, only to trip again, repeatedly slapping their heads against the cement. Others ("17 per cent," said the industry's own welfare review panel) managed to regain their footing, then career down the slope onto the bloody killing floor, where an excruciating game of cat and mouse would be played between slaughterers and the cow.

On our way out of the abattoir Dju points to a big shed. "That's where the stunning is done," he says.

I stop and gesture to my translator. "Wait – there's stunning here?"

He nods. "We hardly use it. Only when there is big demand, like for festivals."

"You mean for Muslim festivals? So stunning isn't anti-halal?"

My translator interprets this for Dju. He listens intently and smiles. "No, stunning is okay, it's halal. It's easier too, but more expensive."

*

"Sure, by our standards the Mark 1 box is no good, but by Indonesian standards it has been a vast improvement. The majority of workers use the boxes properly, even though the boxes meant about five jobs per cow were lost," says Greg Pankhurst.

I understand his dilemma. It's the same dilemma Cameron Hall of Livecorp outlined to the journalist Sarah Ferguson in the *Four Corners* interview. "You've been to Indonesia, you were there," Hall practically pleaded. "It's a developing country."

Later, Pankhurst shows me the new slaughtering box MLA is planning to begin installing. We stand underneath a blue tarpaulin as he points out how it will work. "The cow will enter here, it'll be locked into place with this, and then this will rotate them onto their sides—"

"Hang on," I interrupt, pointing to the rotating mechanism. "How will that work?"

A trace of a smile appears on his face. He knows what I'm thinking. "Hydraulics," he says.

"Hydraulics? *Hydraulics?*" I repeat. "How the hell is that going to be maintained *here?*"

I give the hydraulics a month, two months maximum. Greg Pankhurst isn't even going to give them that. "We've developed our own adaptations to the Mark 1 box," he explains, welding in a manual "squeeze" that locks the cows in place, so a worker can easily stun them from above. Then the unconscious animals will trip onto the floor as before, but without an inkling of awareness.

Since the ban, Pankhurst has announced to all his buyers that his cattle will only go to abattoirs that stun, and has invested in $30,000 of stunning equipment for even the smallest slaughterhouses. "We can't deprive people just because there are too few animals to stun.

"I've had Indonesian abattoirs say to me, 'But I bought the cattle, I own this cattle now, you can't tell me what to do with it.' And so in these cases, we've installed the stunner, taught them how to use it, upgraded their ramps and facilities, paid for the electricity, and revisited on a number of occasions. And if, even then, they haven't budged, and still won't use the equipment, I've had to say, 'Well, okay, we're not selling to you anymore.'"

"Where has that happened?" I ask.

"There are places like Bandah Aceh, which is fundamentalist Muslim. We won't be sending cattle there, because they won't stun, they say it is not halal. They may agree to use the new boxes, in which case other exporters may send cattle to them."

When Greg and his Indonesian business partner broke the news to one such slaughterhouse in Sumatra's north – that they were cutting off their supply – things got tense. "We were standing on the killing floor, explaining why they wouldn't be getting our cattle anymore and then one of the guys put the hose on me. Dickie, my partner, said we'd better get the hell out of there."

Often, after the cows' throats are cut in Indonesia, the slaughtermen put the hose on them. That night, in Medan, it was clear how the workers felt about Greg.

*

"Have you ever seen animals being killed?" Greg asks me.

I'm about to reply, "Of course I have," when I suddenly wonder, *have I?* I feel like I have, but I go blank, unable to recall for certain. Perhaps I've only seen it on film, or in photos.

It's around 10 p.m. when Greg picks me up. We're in Lampung, on the island of Sumatra. During the day, he has shown me around his feedlot, where each year 80,000-odd head of Australian cattle arrive from the Northern Territory to fatten for two to three months. It is nothing like the government-owned Dharma Jaya lot in Jakarta. Clean and finely mulched coconut husk is strewn underfoot daily, the cows have large troughs of clear water, and tractors constantly run up and down the aisles refilling the feed boxes. Unlike in Australian feedlots, where cattle are fed grain that is especially grown for them, Pankhurst's cows are fed mostly by-products from plantations, such as pineapple, tapioca and coconut.

Over a thousand locals are employed there, including 700 local women, who plant and farm small areas of the feedlot by hand. Smiling, in sing-song voices they call out "Hello, Mister Greg!" At around 4 p.m., trucks and buyers arrive at the mouth of the feedlot and buy, in total, 250-odd cows, which are then taken to abattoirs to be killed that night. The selected cows are weighed and the microchips in their ears scanned, as part of Australia's new regulations post-ban requiring exporters to track all cows to slaughter. It is hoped these tracking devices will ensure international animal welfare guidelines – designed by the World Organization for Animal Health not to represent "best practice," but to provide a baseline, a starting point, for developing countries that don't have any animal welfare arrangements in place – are followed. Note that by these standards stunning is not mandatory.

Now, driving down the curve of a winding street, we pull into a driveway. It's a black-painted night except for the rectangular light of the abattoir, which, if it didn't have its roller door wide open, I would have assumed was someone's two or three-car garage.

Amid the bellyaching of frogs, I can hear the whisk-whisk-whisk of knives, men and boys getting ready, their blades being looped, as if led by an invisible needle and thread, over sharpening stones. I can't shake the feeling that these skinny young Indonesian workers stealing glances at me, hiding smiles under their hands, are wondering why Australian women, be they Animal Australia's Lyn White, the ABC's Sarah Ferguson and now me, are so obsessed with watching them kill animals.

Tonight, nineteen cows are to be killed. I can't see them, not yet, but I can hear them, hooves gently stomping out the smell of coconut husk. They've been put in a blacked-out section, behind the killing floor. Greg introduces me to the owner of the small abattoir, a man sitting at a desk facing the area where the cows are to be stunned, cut, dragged and portioned. Then I meet two of Greg's specially employed animal welfare officers. "We have employed forty-odd new permanent staff," Greg had told me earlier. "To monitor the abattoirs, to scan the electronic tags, to educate and regulate animal welfare."

"So that's new jobs?" I'd asked.

"Yes."

"And higher skilled jobs too?"

"Yes."

"Well, there's a plus, surely?"

We had been talking about the loss of jobs the higher standards would mean. First, there'd be the abattoirs that would have to close because they'd fail to change their habits, and second, there's the reality that stunners and better slaughter boxes would mean fewer hands on deck.

I shake the hands of Greg's new welfare officers. One of them shows me how his scanner works, how he will oversee and scan the death of every animal tonight, and upload his records to a database.

Then the killing begins.

On metal catwalks over the cows, two workers crawl and shimmy, using the soft-clothed heads of mops to nudge a cow forward, separating it from the others into a curved corridor, lightly pushing it along, accompanied by the hiss of the stunner's compressor getting ready and the minor thumps of trapdoors until the cow is well and truly removed from its herd.

The last nudge sees the cow inside Greg's adapted Mark 1 box. A man stands above it, pulls the head restraint, and the cow is now locked in. And then the man brings the stunner down, onto the top of its head. I lean, crane my neck, trying to see, but all I see is the metal box, and then, phhhhh-thump, like the sound of a nail gun, and it is done.

The cow falls, I still can't see it, but the box inflates and deflates as the cow loses consciousness, the rubber around the skirting flapping. Then the door opens, and out she falls, tripped, floppy.

A slaughterer steps forward, pinching the loose skin around the Brahman's neck, holds it taut and cuts. Blood pours out like grain from a cut sack. The eyes go first. The slaughterer steps away and workers kneel around the cow, holding it firm, their palms flat, gentle and confident on its leathery flanks, while life ripples through the animal. I never expected it to be beautiful.

Her legs start to move, as if swimming, her nerves doing a final shimmery dance, and then comes a low growl, her body moves, rolls slightly, and the men re-adjust their hands, and then, finally, the air is expelled, past her vocal cords, giving them one last tinkle, like wind chimes.

I get a slight panic. Maybe she's not dead.

"Is she dead?" I ask Greg, trying to hide the worry in my voice, not wanting him to regret bringing me here. "Are you sure she's dead?"

He looks at me quizzically. "Yeah, she's dead. Dead on the first cut."

When the blood slows, the cow's open throat drained, the men and boys cut her head off. The eyes are filmy blue. One lifts the head by the ear, his skinny arm flexing, and drops the head in the corner with a whump.

The workers have separated into three cutting groups, each with a number of carcasses to prepare for their bosses at the market. They drag the cow off the concrete slope, away from the metal box and across the killing floor. Then they balance it in a wide groove and cut a line down its middle, like a bolt of material. Its skin peeled back, belly up, guts splayed, limbs akimbo, the cow could be anyone. It could be a large man. The workers unpack the organs and then, standing barefoot in the ribcage, begin cutting lengths of meat. Back along the catwalk, the mop dangles, hooves shuffle, the compressor hisses, doors clang, until, like a pinball, another cow is lined up.

One, two, three cows. Stun, cut, hold, cut, drag and cut. I stand close to the metal cage now, out of the cows' line of sight, but so I can watch the captive bolt, the shock instantly turning the animal off, the legs going loose and then dropping, out of sight.

Then four. Four doesn't work. It's a breach. The stunner misses and the worker in charge of opening the door to the box doesn't check. He clangs it open and instead of rolling out on its side, the cow is on her feet, staring out at the killing floor. For a moment, everyone stares at the cow, and she stares at us.

Then half the workers in the room run backwards, out of the abattoir, one quickly pulling the roller door across. Greg and I follow, stand outside, watching through a thirty-centimetre gap. The cow slips down the slope, skitters onto the wet floor, the grates mucking up her balance, hooves getting stuck and she's skating on blood and water, the undone bodies of cows all around her.

But she's not bellowing. She's been dazed a little. The stunner must have just clipped her. Five men are left in the room, they dance around her, one gets a length of rope, they try to lasso her legs together, try to get her on her side, but she keeps getting up, legs then slipping out from underneath her before they can draw the rope tight, and the men, the boys, they're just bits of reed in her way. They jump around, saying, "Whoa, whoa, whoa."

"Whoa, whoa, whoa," and she's starting to make sounds, her eyes darting, like she's catching on to this whole ruse, and all I can think is *just kill her, just kill her, just fucking kill her quick as you can, don't let this play out*, and I know Greg is thinking the same thing. Then the workers start trying to push her back up the concrete slope, back into the box, where the stunner is, an impossible task, but they know I'm here – a journalist is here – and one of them is saying, "We stun, we stun" and I suddenly realise there is no intuition to this scenario, that animal welfare requires a kind of intuition.

Greg starts yelling at them in Indonesian. I don't know what he's saying. I can suddenly imagine why the men in the *Four Corners* footage slashed the animal's tendons: how else do you get the animal down? There's nothing malicious or brutish going on here; it's more like trying to catch a bird in a house. But this is going to be ugly. The cow is flailing now, she is rearing up, panicking, misreading every move they make, or perhaps reading the situation too well, that it is a set-up, and everyone is going "Whoa, whoa, whoa," and I'm here, they're all thinking, *a journalist is here,* and all I want is someone to be a goddamn cow whisperer.

*

In the morning, Greg picks me up from where I'm staying, a poky room where when I spat my mouthful of toothpaste and bottled water into the sink, it went down the drain and came out the other end, landing on my shoes. As we head to the wet market where last night's cows are being sold, he gives me a pair of plastic crocs: "Believe me, you'll be needing these." Driving past roadside stores selling silver domes, those shimmery dollops you see on the top of mosques, Greg tells me he was furious last night. That after I left, he drove back to the abattoir and went through them. Stayed there till 4 a.m., until the last cow was killed.

I feel bad for Greg. He's let me in, and now he's wondering if he'll pay the price for transparency.

At the wet market, we walk past crates of tofu and fish balls, the ping-ping of machete-cut chicken heads. A seller rushes up to us, thrusting a bag of spaghetti. "Mr Greg" greets people he knows. It is quite wonderful to watch as fluent Indonesian spills out of him, a big ginger-headed, freckle-flecked man from Queensland.

We stop at one beef seller, sawing a section off a ribcage, and as they chat in Indonesian, I look at the cow head on the ground, eyes blue, flies buzzing around it. I feel like I recognise it. I want to lean down, say, "Hello, aren't you from last night?"

"Too much fat," Greg tells me as we walk away. "He was complaining that there was too much fat on the carcass." Indonesians aren't concerned about different cuts like Australians are, but they don't like fatty meat.

Then we troop up a stairwell to the beef section, an open-air zone of white-tiled slabs, hooks and scales, bloody floor. The air is putrid. I adjust my scarf so it falls over my nose. Cows' faces are turned inside out like rubber masks, and in the far corner a small pile of rubbish burns. On top of one slab, a man is curled up and lightly snoring. Kneeling on the cement ground, another man is cutting up a flank, a rough-looking kitten watching him intently. He steps around the small cat without looking at it, letting it grab scraps, lope around and look at things.

The butcher and the kitten, there's something here, something I can barely put my finger on, and perhaps I'm just desperate to find an inter-species relationship to admire. But this unusual agency of an ordinary cat, its freedom to exist, to not be shooed, to not have a fuss kicked up over its presence, is a tiny moment that shines in relief, this thoughtlessly shared space between butcher and cat.

Back at Greg's house, we have a coffee on his balcony. On a clear day, you can see the Krakatoa volcano, he tells me. For more than twenty years he has worked here, and after spending the first ten years in this house with his wife and two children, he now stays intermittently. He recalls his first year working in Indonesia when he was twenty-two years old, fresh out of agriculture school, as hellish and lonely. "It was only the fear of failing that kept me here."

It seems that utilitarian conundrums are a constant. Once, after he moved into this house, a man knocked on the door and offered to sell him a tiger cub. Its mother had been shot and he now had two cubs to sell. At the door, Greg had to weigh it up quickly. "Buy and rescue the cub and risk the man thinking he can make a living out of this, or shut the door, hope to hell that no one else buys the cub and try to forget that it will probably die."

He shut the door.

A woman at Jakarta Animal Aid Network tells me of similar compro-mises, of decisions she'd never consider making back in her homeland, Denmark. "I mean, I'm anti-zoo," she tells me. "But you can't be anti-zoo here and make a difference." She shows me photos of the animals she has seen on a recent visit to the zoo. A brown bear, in a cage in which it can only crouch, is covered in whitish scabs, a fungal infection. In the pelican cage, the birds are stacked on top of one another. Staff don't go in, she says, they just throw food through the wire. Among the critically ill animals at this particular zoo are a koala, a Sumatran tiger and the brown bear. Most are suffering from pneumonia and malnutrition.

The argument that Australia should stay in the live export cattle and sheep trade because it stands a chance of improving animal welfare in countries where such practices are non-existent is part of this conundrum. On *Four Corners*, Sarah Ferguson's interview with Rohan Sullivan, a cattle exporter and the president of the Northern Territory's Cattlemen's Association, summed it up.

> ROHAN SULLIVAN: We need to be moving towards stunning as … our ultimate goal, but recognising that … well it's going to take time and that … we need to be a bit patient about it because there's lots of reasons why stunning is not going to be taken up straightaway.
> SARAH FERGUSON: You say we've got to have patience, but why should the animals suffer while we help Indonesia get its act together on stunning?
> ROHAN SULLIVAN: Because, I think that …

Sullivan pauses. He is silent for a long time.

> SARAH FERGUSON: It's a tough question, isn't it?
> ROHAN SULLIVAN: Yes, it is.

*

"So has the World Society for the Protection of Animals given any long-term thought to perhaps starting up a world society for the protection of women and the appalling things that happen to women in some of these countries?"

This was one of the first questions put to a female WSPCA campaigner in a Senate inquiry set up after the *Four Corners* report to investigate alleged animal cruelty and industry negligence in live export. Baffled, the campaigner replied that to her mind the two issues were not "mutually exclusive," before trying to steer the wayward questioner, Liberal Senator Bill Heffernan, chair of the inquiry, back to the topic at hand. Heffernan's point was that to be pro-animal is to be anti-human. And indeed, the tide of revulsion in response to the footage of the horrible deaths of Australian cattle certainly mystified many local human-rights campaigners.

The ban raised more than a few questions about the state of Australia's moral compass. Around the same time, the federal government was busy promoting its "Malaysia Solution," a kind of exchange program whereby Australia would redirect 800 boat people to Malaysia – a country with reportedly substandard treatment of asylum seekers – swapping them for 4000 refugees. During the outcry over the cattle trade, one bewildered human rights campaigner said to me, "What about the people coming the other way?" and wryly mused about "ships passing in the night" filled with Australian cattle and asylum seekers.

Others played the starvation card. "It's deplorable to put the welfare of cattle above the welfare of people," wrote Alan Oxley, the chairman of the APEC Study Centre at Monash University, in the *Age*. "Australian cattle account for more than a third of the meat consumed in Indonesia, a country in which nearly 40 million people live below the poverty line. Surely it is more inhumane to use the leverage of inducing food shortages among the poor than to see animals mistreated."

Bob Katter, independent MP for a northern Queensland cattle-country electorate, while condemning the mistreatment of cattle, attacked the Greens' push to ban permanently the live export of animals. "Eighty million out of 240 million people in Indonesia go to bed hungry every night," said Katter, before adding that the Greens and other environmental groups were "evil" for proposing policies that would jeopardise Indonesia's food security. "We have to assume [the Greens'] solution is to starve a lot of people to death."

In the *Sydney Morning Herald*, Dr Jeff Neilson, a geography lecturer at the University of Sydney, wrote that the average Indonesian response to the cattle ban was one of incredulity. To many, "Australia's claims of moral superiority have an awkwardly hollow ring. Animal rights," he continued:

> are unlikely to emerge as a priority in a country where 36 per cent
> of the population is considered food-insecure. Rising food costs as
> a consequence of the export ban in the lead-up to Ramadan will
> trigger social concerns there that would be politically unacceptable
> in a comparable Australian context.

Although he intimated that the Australian cattle ban could give rise to hunger in Indonesia, Neilson was careful not to state this outright. After all, it is nonsense.

This is not to say the cattle trade does not indirectly benefit local Indonesians: thousands are employed to work on feedlots, provide fodder for animals, as well as transport them and so forth. But of all the impoverished locals, the only ones eating Australian beef are the slaughtermen themselves, who might accept a portion of meat as payment for a night's work.

It is well known within the industry that the "protein trade" – as some exporters like to refer to Australia's global meat trade – rarely, if ever, reaches the Indonesian poor. On MLA's feedback channel on YouTube, the agency states that "only the wealthiest 10 per cent of Indonesians currently buy beef, although this is still 23 million people," with Indonesia relying on Australia for less than a quarter of its beef supply.

"Indonesians don't have fridges," became another pro-live animal export catch-cry. They *need* their meat freshly slaughtered. Peter Alford, the *Australian*'s Jakarta correspondent, quickly plugged that one: "Actually, 60 per cent of households have at least one refrigerator, according to the Indonesian Association of Electronics Entrepreneurs."

The fallacies spun on. To prove his point about the Indonesian response to the cattle ban, Neilson referred to a popular blog by Sofyan Arief Ardiansyah, its tagline reading "Aussie: *Jangan sakiti sapiku!* Indonesia:

silahkan perkosa TKW-ku" — "Australia: don't hurt my cows! Indonesia: Go ahead and rape my overseas workers."

Curiously, as the Australian cattle ban was being put into effect, the Indonesian government was being pressured to consider a ban of its own: a moratorium on sending domestic workers to Saudi Arabia, where over two decades thousands of Indonesians have alleged abuse, mistreatment and exploitation. Ardiansyah, as Neilson points out, is incredulous at both countries' attitudes.

But it's as if he did not read on. Of the parallel bans, Ardiansyah wrote:

> How great is the Australian government's concern about the fate
> of his cows ... Australia immediately stops sending hundreds of
> thousands of cattle. No matter that it is a business worth hundreds
> of millions of dollars ... When cows are so protected, we can imag-
> ine the protection afforded to citizens of State. So far, Indonesia
> does not even protect its citizens.

Incredulous, yes — but not without admiration for Australia's level of empathy.

Both Oxley and Neilson go on to reflect on the arrogance of trying to foist our standards on another country. Australia's dealings with Indonesia, says Neilson, reveal "an uncritical neo-colonial projection of Australian moral values into the international arena without due recognition of the cultural and political contexts."

Yet while insisting on standards of animal welfare and slaughter is considered "neo-colonial," campaigns such as Meat and Livestock Australia's "Marketing red meat around the globe" slip under the radar. In Asia, where white meat, such as chicken, pork and fish, is largely preferred (and, if we zoom out to look at the bigger picture, produces barely measurable methane emissions), these local diets are undermined by promotional campaigns claiming that the Western diet is "better for you."

In Japan, the "Iron Beauty" mission set up by Meat and Livestock Australia aims to promote the importance of beef to Japanese women's health and beauty. In Moscow, Russians are treated to Aussie barbecues. In South Korea, young teenagers wearing "Kids Love Beef" T-shirts perform a drumming act on stage to celebrate Australia Day, while the same MLA-spun slogan is printed on Korean delivery trucks. In more than twenty countries, from Cambodia to Mauritius to Jordan, MLA runs culinary challenges using Australian meat. In Indonesia, celebrity pop and TV stars cook MLA-supplied hamburgers while the entrances of supermarkets are decorated with Aussie meat posters – "Beef makes you tall," "Beef makes you strong" and "Beef makes you smart." Advertisements show images of Indonesian families sitting down in a dining room to a Western-style meal, while at wet markets, because imported cattle cannot be branded Australian by law as the cows are "finished" onshore, MLA has come up with "*Daging Sapi Kita*," meaning "Our Beef," so consumers can make the connection between the product and the marketing.

Such overseas campaigns have been heavily amped up as Australian interest in red meat wanes. Here, consumption of beef peaked in the late '70s, when Australians ate *seventy kilograms* per person each year (current Indonesian consumption is two kilograms per person if averaged out). But today Australians eat less than half that, while lamb and mutton consumption has diminished by three-quarters in the past few decades.

*

In Indonesia, we do have a responsibility.

We are now, officially, in a bind.

We can't just stop the trade and say, "We won't do it anymore." We've created the demand. I'm not sure that we ought to have started the trade in the first place, but now that we have, we can't just stop. Or can we? Will Indonesians simply make do with less beef? Or will the Indonesian government let live Indian or Brazilian cattle in, risking foot-and-mouth disease and Australia's own multi-million-dollar program against it in the region?

Or worse, do we risk provoking the ire of Indonesia, cementing its desire for self-sufficiency and control, which will most likely mean increasing its herds and grazing land to satisfy the demand we have created, furthering the destruction of rainforest, crucial orangutan habitat and traditional homelands, areas already being devastated by palm-oil companies?

Are we better off staying in the game and educating developing countries about animal welfare, while at the same time assisting them to introduce industrialised methods of factory farming, a practice that we ourselves are increasingly finding abhorrent?

And finally, do we trust the live export industry, represented by MLA and Livecorp, financially backed by the Australian government, to meet international welfare standards?

Greg, I trust. But Greg is one man and industries are not made up of individuals.

How can you trust the gaze, a gaze that is made up of numbers and units, not living creatures, how can you trust that gaze to truly *see* what is in front of it?

"We did not see any of the cruelty that was shown on the ABC *Four Corners* program," said Ivan Caple, a former dean of veterinary science at the University of Melbourne and a member of the Livestock Export Standards Advisory Committee. Speaking to the senate inquiry, Caple continued, "If we had seen it, we would have stated that. We would have rung the minister."

But Caple, who chaired the MLA and Livecorp–commissioned welfare review of Indonesia's slaughterhouses, and his panel *did* see similar slaughtering practices to those that *Four Corners* revealed to a shocked Australian public. And they *did* state as much in their report to MLA and Livecorp, which was later forwarded to the Australian government. Their report revealed cattle being "rolled," "head slapping," "unnecessary stimulation involving interference with the eyes and tail twisting" and examples of poor throat-cutting – one cow was cut an estimated eighteen times – that may have resulted in "extended consciousness."

The panel concluded that animal welfare in Indonesia was "generally good."

Sarah Ferguson asked Caple, "How does good welfare square with the eighteen cuts then?"

Caple responded that eighteen cuts was the "extreme." The panel, it turned out, had watched in total twenty-nine cows being slaughtered in eleven abattoirs, and in that snapshot they saw the "extreme."

The truth is that in Indonesia everyone saw the same thing.

Meat and Livestock Australia, Livecorp, the industry-commissioned welfare review panel, some cattle importers and exporters, Animals Australia activists, the ABC journalists, the RSPCA and now much of the Australian public all saw the same thing. And yet, the interpretations varied dramatically. The question is, who is out of step with whom?

Is the industry and its internal animal welfare regulators out of step with the public? Or is the public out of step with reality? Following the *Four Corners* report, the columnist Michael R. James wrote in *Crikey*:

> there was something slightly obscene about a majority of Australians' – bogans and latte sippers united – reaction to the clear brutality in some Indonesian slaughterhouses. With the carelessness typical of Australians when it is a matter a million miles from their own privileged lives, they will venture an absolutist "solution" to simply close the whole trade down, not only injuring relations with the largest Islamic nation in the world (also the 4th largest nation in the world) and our closest neighbour (and where about 1.7 million of those same bogans and latte sippers go for their cheap holidays though in the non-Islamic part, Bali, where no doubt they will continue to happily consume their beef rendang) but most obviously not solving the problem at all.

Though snide (why is it that so many columnists seem only able to write if they "get their bitch out"?), James has a point. It was easy to condemn the slaughtering of Northern Territory cattle in Indonesia. It was them – not us. But how do you marry that conclusion with recent undercover footage taken in British abattoirs that reveals a worker kicking pigs as they squirm around his legs, their hooves slipping in blood, struggling to get out of the room, squealing. With electric tongs, the man shocks them, on the snout, on their legs, in their ears. Another worker hits a pig in the face with a metal hook.

Closer to home, in Victoria's Gippsland, an abattoir was shut down last November after footage showed workers using the stunners incorrectly, again all over the body, including in the eyes, and one man letting a panicking pig run straight into a scalding bath.

Or how about the Australian processing of bobby calves? Male calves less than two weeks old are stacked into trucks and taken to abattoirs. In one piece of footage, a driver stands at the back of his ute, hands on hips, talking to two other men. Every now and then he reaches up to his truck to grab and hurl a tentative calf not following the others into the corral, then turns back to the men to pick up the conversation. None of them seems to notice as the black-and-white calves, still learning to walk, some not quite managing it, crawl past on their knees into the abattoir.

In Australia, half a billion animals are raised for food production, the majority in intensive or "factory" farms. Broiler chickens "roam freely" in long corrugated tin sheds. Well, they can shuffle the first step of the waltz for a week at most, but then, as their specially bred bodies grow to 1.6 kilograms in less than half the time of their ancestors in the 1950s, not only do their bodies inflate and fill the space, but their legs can't hold them up, slowly rotting in the ammonia of their own shit. Add to this the climate-controlled permanent night of their forty-two days on earth, as they're tricked into a night-time routine of not moving around, not burning calories.

The condemnation of Indonesia's slaughtering practices was valid – but how valid? How is it that Australians could not see their own reflections in that footage? Many self-described animal lovers confided to me that they wanted to go over to Indonesia and beat the workers after they saw the footage. Sure, local butchers did report a temporary drop in beef sales in the first weeks after the Four Corners report, but after a month or so, what with the ban in place and all, sales picked up again as we managed to get our ghosts under control. For the most part, Australians empathised with the cows, not the workers – and that is truly an irony.

On the operating table lay a baboon, its front cut open and kidneys carefully removed. The surgical team closely monitored the animal's pulse. The primate had been sent, along with thirty-odd others, from Sydney's baboon colony, a colony specially bred for scientific research under the auspices of the Royal Prince Alfred hospital. A surgeon prepared to transplant a genetically modified pig's kidney into the baboon.

"It was unpleasant," says the same surgeon, now retired, as he sits opposite me in a café on Lygon Street, a busy Melbourne street filled with Italian restaurants. "I had to keep the motivation for the study at the forefront of my mind."

"Which was?" I ask.

He looks at me, slightly surprised that I need to ask. "For the greater good of suffering humans."

"And did it work? Reminding yourself of the bigger picture?"

The surgeon shrugs a little. "Surgery, I feel, is a privilege. A privilege to see inside of humans, to see inside of these creatures, to be so close to baboons. But it felt wrong to disturb it." He tells me that as he has gotten older, he has softened. "I don't even go fishing anymore."

This surgery is xenotransplantation, *xenos* meaning "foreign" in Greek. Twenty years ago, in 1992, the creation of the first transgenic pig, "Astrid," whose organs carried human genes to stop the human body rejecting them, made headlines around the world. The once sci-fi concept that every human could have their own pig, a kind of living suitcase of spare parts, was suddenly becoming real. Of course, human trials were still a long way off. This was where the baboons came in.

The surgeon's study was a success, to a degree. The hyper-acute rejection that generally occurs within minutes of inter-species transplants had been avoided, but then, four days later, the baboons developed coagulation defects. "It's always the way: once you solve one problem, another presents itself."

*

"What comes in, never comes out," an animal technician says to me when I ask if a laboratory animal has ever gotten loose in the corridors of the busy Melbourne hospital in which they're kept. My childish mind imagines chaos as rabbits and rats and mice dart under the feet of patients pushing IV drips.

"Here, I'm the voice of the animals," she tells me, describing her work as a kind of mercy – granting a creature leave in the face of the inevitable. "If I see a mouse or a rat suffering, I'll tell the scientist to cull it," she continues. "A couple might say to me, 'No, a few more days and then I'll cull it,' but I get the final say. I'll tell them to cull it on the spot." Before working here, this technician worked with macaque monkeys at a Monash University research facility and with students studying medicine, practising surgery on piglets. "The macaques had electrodes implanted in their heads," she explains, "and the researchers recorded their brainwaves while they did activities, such as computer games and puzzles."

Often the animal technicians, who are responsible for the laboratory animals' wellbeing, are the ones who have a connection to them, while most scientists cultivate distance, or at least the appearance of distance. When I ask the surgeon if he ever mentioned his uneasiness about working on baboons to his colleagues, he quickly shakes his head: "No, we never spoke about it."

When the technician tried to explain to researchers that she had a bond with one of the macaques, that he took food directly from her hand through the wire mesh, they didn't believe her. "There was one day," she tells me, "I was having a bad day, and was having a cry." She was sitting on the floor of her work area, her back leaning against the macaques' cage. "And suddenly he was there, putting his fingers through the wire to touch me."

It's not just the primates that may present problems for these workers if they look at them sideways, briefly losing sight of their scientific motivations. "In the early '90s, I did a series of pancreas transplants in dogs," says the surgeon. "I found it very difficult. It was the first time I thought, *I'm not sure I want to do this.*" He pauses, then explains, "Dogs have a very different order of response." The dogs kept trying to connect with the researchers, and sometimes succeeded.

During an afternoon I spent with the Australian writer Robyn Davidson, I discovered that she worked briefly as an animal technician in a Brisbane laboratory in the mid-'70s, when she was a young adult. "We literally lived with the animals," she said, "while the researchers just turned up every now and then to update their results. So they weren't there when one of the dogs, with this battery-powered device permanently fixed on its neck that released a fluid, or something, that went straight to its heart every ten hours – they weren't there when, every ten hours, the dog would …" Robyn couldn't find the words; instead, she hunched low over the table between us, her hands splayed and neck arched, back rigid, and howled, a long blood-curdling howl.

"Oh god, what did you do?" I said, my hand over my eyes.

"I killed her. Put an injection into her paw and I let the other dogs go. There was one puppy, and I took him with me."

*

In December last year, the US National Institutes of Health (NIH) announced that it would dramatically limit the use of chimpanzees in medical research after an expert panel recommended the use of this particular primate – humanity's closest relative, alongside the bonobo – be curtailed after finding most experiments had feasible alternatives. In its report, the panel said the use of chimpanzees should be reserved only for studies where no suitable alternative was available or where testing on people would be unethical, and only for life-threatening or debilitating conditions. "We found very few cases that satisfy these criteria," said Jeffrey Kahn, who chaired the panel.

In 2009, public opposition to such testing was fuelled when footage filmed at the largest primate-testing facility in the United States was secretly obtained and played on national television. A Humane Society investigator had gone undercover and worked at the facility in Louisiana

– where almost 5000 primates are housed, including 347 chimpanzees –
for nine months. Many animals were simply "warehoused" and bred for
potential research uses. However, when it came to those used for testing,
the footage was grim.

The setting recalls a standard "Cell Block A" from television. Workers
in biohazard suits and masks walk down corridors lined with cells, and
at the front of some cages chimpanzees jump, hoot and shake their bars;
in others they scramble to the furthest corner as if to make themselves
invisible.

Some chimps barely register the workers' presence, huddled in the
dark, head in black palms. The seriously traumatised chimps are kept in
solitary confinement, while in the smaller cages macaques frenetically
backflip and walk in circles.

Stopping at a cell, one staff member points a tranquiliser gun at a chosen
chimp and it starts to scream. Sitting on a perch at the back of the cage, it
falls onto the cement floor below with a thud when the dart takes effect.

Footage from inside the laboratory shows four monkeys anaesthetised
and casually stacked, their legs hanging over the edge of a bench. One
slides off and hits the floor.

A baby macaque is restrained on the surgery table. It looks at the people
holding it down, then peers curiously at the syringe entering its abdomen
before wrenching its head back in a yowl.

In another scene, a worker whacks a monkey in the mouth three times
with a pole, trying to make it open its mouth.

An older male chimp, used for testing for decades, becomes fascinated
with the investigator. The footage shows him gesturing and making hand
signals at her. She discovers later he had been taught some sign language.

In the New York Times last year, Roscoe G. Bartlett, a Republican congress-
man and former physiologist at the US Navy's School of Aviation Medicine,
wrote of his work with primates in the fifties, engineering and testing
respiratory support devices on monkeys before sending them into space.
"At the time," he wrote,

I believed such research was worth the pain inflicted on the animals. But in the years since, our understanding of its effect on primates, as well as alternatives to it, have made great strides, to the point where I no longer believe such experiments make sense – scientifically, financially or ethically.

Adding fuel to the Great Ape Protection campaign, a network of newspapers owned by the McClatchy group across fifteen US states ran a series of articles on the history of testing on chimpanzees, after it obtained access to records that an advocacy group had finally won after a five-year legal stoush with the NIH. The McClatchy reporter, Chris Adams, settled on the story of Lennie, a chimpanzee captured in Africa in 1962 and strapped in a capsule for government test flights, to illustrate the life of a test chimp.

Chimps were used, literally, as crash-test dummies. For example, in Project Abrupt Deceleration, chimps suffered third-degree burns from windblast, while Project Whoosh, an engineering study on "ejection" seats, saw hundreds of chimps fatally ejected from missiles at supersonic speeds.

Jane Goodall, the renowned primatologist, described the famous and largely misunderstood grin of "Ham" – one of the few chimps sent into space – as a "fear grimace." Ham's return to earth had gone a little awry, his capsule landing in the ocean a hundred kilometres from the recovery ship. By the time he was fished out of the sea, water had seeped into the capsule. "Ham was rewarded with an apple," says the New Mexico Museum of Space History, "and joyfully 'greeted' everyone aboard the rescue vessel." Despite his seemingly high spirits, when NASA later held a press conference with the "astro-chimp" and tried to get him back in his space capsule for photos, four men were unable to wrestle the screaming Ham into the contraption, even though he had been conditioned to know that he would be given electric shocks if he disobeyed.

Unusually, Ham was transferred to a zoo after NASA outgrew him. Lennie, however, like many others, was onsold to undergo further research and drug testing. He was subjected to numerous spinal taps and fed bananas laced with triparanol – a drug already removed from the market because it damaged human eyes. In the '70s, he was used as a breeder to increase the supply of lab chimps, and was later infected with viruses such as HIV and hepatitis, undergoing hundreds of blood tests and biopsies. Adams continues the story:

> Lennie's 540-page medical file and "Chimpanzee Resume" detail the regular tests and blood draws, which required a "knockdown," an anesthetizing dart. It was something Lennie resisted. "Animal excited at knockdown," a 1998 record said. "Excited knockdown," a record in 1997 said. "Pyrexia" – fever – "possible due to excitement of sedation," it read in 1998 again.

For all of science's precision, the words "resisted" and "excited" were not even shadows of Lennie's reality.

In the new 2011 guidelines for limited research to be conducted on chimps, the advisory panel recommended studies be limited to "acquiescent" animals, as in chimpanzees trained to "present" for testing. *Give me your arm for a needle and I'll give you a treat.*

The procedure of "knockdown," with its screaming, scrambling, being held down by four to five people wearing masks and biohazard suits, its writhing and physical pain, is undoubtedly traumatic. But this compromise, the promotion of better practice through creating a line of voluntary chimps, seems grossly manipulative, a veneer of choice.

For forty years, Lennie was kept in a small cell and, along with thousands of others, used for testing purposes. Then, in 2002, at a primate facility in the New Mexico desert, he grabbed the side of his cage with all four limbs and died.

A recent series of studies exploring the psychological impact of life in laboratories on chimpanzees revealed that the primates suffered depression, anxiety and post-traumatic stress disorder. In the past, much of the scientific community dismissed such findings as "anthropomorphic." But the data, collected and analysed by the researchers, is difficult to refute.

A comparison of the behaviours of chimpanzees used in experiments with those of wild chimpanzees revealed a high prevalence of mental disorders in the former. Listless chimps obsessively plucked themselves bald and self-mutilated. Others lay in the corner of their cage, knees up, sometimes rocking on their heels, head in hands and refusing to eat.

In *Developmental Psychology*, a medical journal, the story of Billy, a former birthday party entertainer, revealed the extent of the trauma one might find in a captive chimp. Billy had lived and bonded with humans for fifteen years before being sold for medical research, after which he rapidly deteriorated mentally. Used for research into HIV, polio, measles and hepatitis, Billy was subjected to hundreds of "knockdowns." After one such knockdown, the chimp awoke and chewed his thumbs off.

In the 1980s, the number of chimps bred for research nearly doubled in the ongoing quest to find a vaccine for AIDS, but despite being our closest relative, the chimps turned out to be a failure. They could contract the virus, but they were immune to its effects. In 2007, an article published in the *British Medical Journal* concluded, "When it comes to testing HIV vaccines, only humans will do." Research, however, migrated over to the rhesus monkey, which is said to reproduce human HIV more faithfully.

In 2000, in the United States, the *Chimpanzee Health Improvement, Maintenance and Protection (CHIMP) Act* was passed, allotting funds to retire and house surplus lab chimps owned by the government. "Retirement for chimps is, in its way, a perversely natural outcome, which is to say, one that only we, the most cranially endowed of the primates, could have possibly concocted," wrote Charles Siebert in the *New York Times*.

The retirement villages are strangely familiar. In the sleeping quarters, residents have their own running water, hammocks, mirrors, windows and sound systems set up with TV, CD and DVD players, while outside there are children's jungle gyms. In Los Angeles, there is a privately run sanctuary for washed-up chimpanzee actors.

After fourteen years of living in a lab, 29-year-old Billy was transferred to a sanctuary. But despite these surreal twilight years, he never recovered from his days as a test chimp. Up until his death seven years later, he remained agitated and scared. Carers noted that he screamed if the door to his cage was left open, and they had to check the door before he went to sleep at night to make sure it was locked. Billy's ability to mix with other chimpanzees was near zero. The carers summed up his state as follows:

> Tyrant in social settings, not tolerated by most, feared by some, constantly challenged by others. Has had the most violent attacks made on him by the whole group. Great difficulty in any social setting.

*

"Thanks to animal research, they'll be able to protest 23.5 years longer," says a poster from the Foundation for Biomedical Research, a Washington DC organisation set up in the '80s to counteract the animal-rights movement. Accompanying the laconic message is a black-and-white photo of angry young protesters holding placards against animal research. In small print beneath the photograph, the foundation adds, "According to the US Department of Health and Human Services, animal research has helped extend our life expectancy by 23.5 years. Of course, how you choose to spend those extra years is up to you."

On another poster, a photo of a rat sits next to a photo of a girl, her head cocked cutely at the camera. "Who would you RATher see live?" is typed over the two subjects. In other words, it's the rat or her. You decide.

Few other issues inspire more "Philosophy 101" quandaries than animal testing.

When I spoke to scientists, researchers and technicians who had worked with lab animals, few failed to pose the question: "If you had to choose between your — [insert child, parent, sibling, loved one] and a — [insert a baboon, pig, rabbit], what would you do?"

It is a question designed, albeit politely, to shut me up. And it worked, for a time. After all, what do I know about science? Who am I to say what goes on in a laboratory or a hospital? But, eventually, I found my answer.

I'd choose my loved one, of course, but if I were in a better frame of mind, outside of this emotionally charged state in which no dilemma should be resolved, I might wonder why someone was holding me to ransom with a baboon.

What kind of sicko is lining up guinea pigs at the foot of hospital beds and knocking them off on the nod of desperate parents?

The problem with discussing the use of animals in experiments is that it is difficult for things not to devolve into the kind of high-school debate where teams "for" and "against" use tit-for-tat statistics and contrasting quotes from eminent people until invariably the final speaker in the "for" team announces that, for all their ethical airs and shrill moralising, the speakers on the "against" team wouldn't even *exist* if it weren't for animal testing.

It's a comment that cuts to the core of the issue and it's 50 per cent bullshit.

*

Humanity's expectations of health changed dramatically and forever with the development of vaccines and antibiotics. Before disease was linked to germs, and an understanding of microbes and bacteria was acquired, death cast its cloak over entire cities. Its approach was inevitable and the fight, the human struggle to escape, may have felt like a helpless animal's struggle in a trap: humans were being tortured by some omniscient, stronger being.

Then, over a period of two centuries, physicians learnt how to isolate disease-causing microbes, locate their origin and point of transmission, and map their path through the body. Most significantly, they discovered that humanity could not only fight disease, but also could be ready for it. The idea of immunity was born, and now we in developed countries cannot imagine life without it. The dread that once accompanied rabies, cholera, tuberculosis, polio and typhus is largely forgotten. Smallpox, for example, despite killing more than 500 million people worldwide since 1900, has not been seen in the Western world since 1978.

There is no doubt that humanity's elemental understanding of the body is due to studies on animals. By injecting, tracing, testing and mass producing medicines using animals (before in vitro methods, certain vaccines were made from the crushed spinal cords of rabbits and monkeys), humans have been given a new lease on life.

Many historians trace these pioneers of the body back to Galen of Pergamon, a physician of the second century who, after Roman law banned the dissection of human corpses, resorted to cutting up animals, dead and alive. Perhaps his most poignant revelation was that blood flowed through veins, not – as was commonly believed – "pneuma," an esoteric mixture of air and fire. Galen, say many, initiated the conception of the body as mechanical rather than mystical.

From here, physicians and even artists (think of Leonardo da Vinci) dodged the wrath of religion, peering into the corpses of humans and animals. Vivisection was born, the dissecting of live animals, with physicians twanging veins, fiddling with nerves, putting their fingers in still-beating hearts to feel the contractions. Equipped with the Cartesian theory of humanity's mental supremacy – "I think, therefore I am" – they could dismiss the writhing and howling of these live animals under the knife as pure mechanism. Opening up the chest of an animal and poking around was like popping the bonnet on a car. The sounds they made were simply two cogs colliding.

But the ghoulishness, in spite of the accompanying enlightenment, wasn't welcomed in all quarters. Physicians in Europe, mainly France and Germany, were particularly enthusiastic vivisectors, their work often brimming with spectacle, while the physicians in England were tempered by a less than approving public. Since the early 1800s, the United Kingdom has been privy to various animal welfare bills and was the birthplace of the world's first anti-vivisection group and animal welfare society, which with the Queen's blessing later became the RSPCA.

Two schools of thought were both prevailing and colliding, those of René Descartes and Jeremy Bentham, an English utilitarian philosopher not opposed to animal experimentation per se but to the infliction of unnecessary pain. In 1780, Bentham's much-quoted line of enquiry was simply, "The question is not, 'Can they *reason*? nor, Can they *talk*? but, Can they *suffer*?'"

Today, we know that all animals suffer. There is no question that animals are sentient beings. Even fish have been found to feel pain. Lobsters try to climb out of pots of boiling water. But the line between suffering and mechanism no longer seems to matter as much as, say, the question: do they suffer intelligently? Or, put more bluntly, just how human is this animal's suffering?

In Canada, psychologists recently developed the bizarre "mouse grimace scale," their results published in the journal *Nature Methods*. After subjecting mice to different levels of pain – dipping their tails in hot water, inflaming their bladders, clipping tails, damaging and constricting nerves, keeping them alive post-operation without pain relief for two weeks, cauterising and injecting them with various irritants and acids – the researchers then analysed video footage of these acts and revealed that mice, like humans, signal their pain through facial expressions.

As if to demonstrate the gulf between the public and the scientific community, a public outcry forced the university where the mouse grimace scale experiments took place to investigate whether welfare protocol had been followed, while many within the scientific community saw the study as an important harbinger of similar studies to ascertain the grimace scales of other species and, ultimately, to help gauge levels of pain in humans.

Surely these opposing responses reveal a knot at the core of our dealings with animals? The psychologists responsible for the mouse grimace scale were cleared of any wrongdoing: they had followed the animal welfare and ethical procedures set for them by the scientific community. And yet, many outside this closed world thought otherwise. Again, we have a disjunction, this time with an institution and its aligned community of scientists utterly out of step with the public. Or is it vice versa? Is the protesting public, ignorant of the nuances of science, acting unreasonably?

In J.M. Coetzee's *Elizabeth Costello*, the fictional character of the book's title discusses what she believes to be the self-perpetuating power of reason, in a lecture titled "The Lives of Animals." Does reason, asks Elizabeth Costello, allow access to "the secrets of the universe," or is it "the specialism of a rather narrow self-regenerating intellectual tradition whose forte is reasoning ... which for its own motives it tries to install at the centre of the universe?"

Costello tells the story of Sultan, a captive chimpanzee housed in the Prussian Academy of Sciences in the early 1900s. In a series of intelligence tests, Sultan is starved and presented with problem-solving activities, involving tools such as crates, stones and sticks, which he must accomplish if he is to gain access to bananas.

> The bananas are there to make one think, to spur one to the limits of one's thinking. But what must one think? One thinks: Why is he starving me? One thinks: What have I done? Why has he stopped liking me? One thinks: Why does he not want these crates any more?

But, Costello continues, these are the wrong thoughts. In fact, the more complex the thought, the further away is Sultan from getting his bananas. In the end, the captive chimpanzee uses the stick, stacks the crates and unloads the stones – and gets the bananas. By the experimenter's reasoning, Sultan is measurably intelligent. But, says Costello:

> In his deepest being Sultan is not interested in the banana problem. Only the experimenter's single-minded regimentation forces him to concentrate on it. The question that truly occupies him, as it occupies the rat and the cat and every other animal trapped in the hell of the laboratory or the zoo, is: Where is home, and how do I get there?

Three years ago, a 31-year-old chimpanzee at a Swedish zoo was discovered to be collecting stones before the zoo's opening hours and storing them at various locations in his enclosure. He had also learnt to recognise weak parts of the concrete structure and had chipped away parts of it to add to his hidden piles of ammunition. Hours later, when the zoo was crowded with visitors, the chimp would reappear, now furious and agitated, and throw the rocks at people.

The strategising chimp was written up in *Current Biology* as part of a study that aimed to show that certain animals could anticipate a future mental state. The early-morning planning, the secret stash of rocks and the obsessive chipping away at his own structure for projectiles was evidence, concluded a cognitive scientist, that certain animals can plan for future events.

I've no doubt that this particular chimp demonstrated all this, but would it be unreasonable to add that the chimpanzee was also extremely pissed off, bored and didn't want to be in a zoo?

*

In 1914, as the US science writer Daniel Engber recently noted in *Slate* magazine, the chair of the Council on the Defense of Medical Research, Walter Cannon, warned journal editors to excise from their manuscripts any expressions that were "likely to be misunderstood" or turned against them by animal activists. It was the beginning of scientific "newspeak." Starving became "fasting," inducing brain damage became "brain insult," poison became "intoxicant," he or she became "it," and animals became digits, numbers which are, in the case of primates, tattooed across their chests. Engber writes:

> That doublespeak (by now having become a matter of habit) obscures some of the incidental cruelties of animal research. But it hides just as well the attention and care that are essential to working in the lab. An experimental macaque costs about [US]$8000 and may require months or years of training before it can start producing useful data. That is to say, its continued health is of

extraordinary value both to the professor who paid for it and to the graduate student whose dreams of a thesis depend on its wellbeing.

Engber goes on to recount his own animal experiments in the lab:

> I've also experimented on cats – kittens, really – by probing their exposed brains with an electrode to see where tiny shocks might palpitate their feet. (We were studying neuroplasticity and how behavioral training affected the development of the motor map.) I spent time with the animals every day, teaching them to grab morsels of meat from a plastic container with their little paws ... I stroked their bellies, too, and scratched under their chins. But there's no mention of those affections in the published results of the study. (Kittens "were trained to reach through the aperture to grasp the beef from a narrow cylindrical food well (3.2 cm inside diameter; 5 cm deep) using their preferred limb only," we wrote.) Nor did we mention that the animals – some as young as 3 months old – were euthanized at the end of each "intracortical microstimulation" experiment.

Today in Australia, according to statistics collated by the Australian Association of Humane Research, close to 7 million animals are used in research and teaching annually, and the figure is growing, not declining. Not included in the tally are fruit flies, snails and other insects, while a large majority are purpose-bred fish, mice and rats, live creatures which often come with patents, can be ordered over the phone and are sold much like oranges and apples, but with more paperwork.

Other animals, such as dogs, cats, birds, cows, sheep, pigs, possums, lizards and rabbits, appear in laboratories every year in their thousands. Cats and dogs are purpose-bred or sourced from approved pounds, while primates are imported (on the Indonesian tarmac, planes are loaded up with an annual maximum of 15,000 monkeys and flown around the world) or sourced from baboon, marmoset and macaque colonies in New South Wales and Victoria.

It is near impossible to gain access to these colonies. After I sent an enquiring email to a head researcher at Sydney's baboon colony, he accidentally pressed "reply all" and I was privy to his message to a superior. "Just letting you know I received this," he had written. "Obviously I won't be responding."

Security is tight when it comes to using animals in research. This, say scientists, is due to attacks by animal-rights groups and threats to staff that have culminated in vandalism and the ransacking of laboratories. In the course of my enquiries, a scientist friend of mine working at a university told me that her department had been warned to keep an eye out for a journalist who was snooping around the campus asking questions about the university's use of animals. "Is it you?" she asked.

"I don't think so," I responded. At least, I don't recognise myself in that description.

"All projects have to go before in-house scientific and ethical committees," says the retired surgeon. "Their examinations of our reasons to use animals are vigorous. You have to justify what you're doing, prove you'll have vets and animal technicians caring for the welfare of the animals. You have to prove that it's necessary to use animals and ensure the animals won't suffer."

"You mean ensure you'll ease their suffering?" I ask.

"No, that they won't suffer. Just as you would with humans."

I don't want to be annoying – this surgeon has not agreed to meet with me so I can pull apart the way he speaks – so I bite my tongue. But the language bothers me. I am a wordsmith, after all. I'm not saying the reason for an experiment is valid or invalid, but to speak of another's suffering as if it's something you can control – something in me curdles at the thought.

*

It is difficult to pinpoint exactly when animals in the laboratory went from being indispensable to disposable objects. There have always been cowboys, plucking the veins of a live dog just for the heck of it, alongside the serious practitioners. And herein lies the problem of animal testing, the quandary of necessity.

Let's go back to the high-school debate. The final speaker in the "for" team has just declared that the "against" team wouldn't exist if not for animal testing. The last speaker for the "against" team is yet to respond.

She is nervous and would desperately like a puff on her asthma inhaler, but is savvy enough to know that would signal defeat. Glancing at her palm-sized prompt cards, she cringes a little at their naivety, her childish and bold handwriting. Her two fellow teammates are manically scribbling and pushing ineligible notes to her.

She takes a breath, stands and nods at the Chair, who in turn checks his watch. Then, in a voice which sounds foreign to her, polished like a pollie's with only a slight waver at the end of each sentence betraying her, she begins.

"Yes, as the last speaker said, humanity now has immunity. We enjoy a bill of health that no other generation of our species has enjoyed before. And for that, we have animals and great scientists to thank. But, I ask the team for testing, are we immune to commercial imperatives? A vast industry has been built up around the laboratory animal, and it's difficult to work out what now comes first – the disease or the product.

"Between 1959 and 1975, the global use of animals in laboratories more than doubled. In *Of Mice, Models and Men: A Critical Evaluation of Animal Research*, Andrew Rowan, a biochemist and the president of Humane Research International, states that animals used in tests increased world-wide from 100 million to 225 million in that sixteen-year period. Today, it is estimated that 100 million animals are used *each year*, worldwide. For each new drug, over 500 ani—"

The speaker looks up. The eyes of her audience are glazing over. Quickly she flips past the other statistics she was planning to recite. Pushing down on her panic like clothes in a suitcase, she tries to think ahead, before continuing:

"Animals have gone from objects in the study of fatal diseases to 'necessary animal models' in sub-standard career-climbing thesis topics, to studies for livestock efficiencies such as how to create a better 'meat to bone' ratio, and, most significantly, to product testers.

"Cleaning detergents, pesticides, soft-drink flavour enhancers and studies on the effects of black tea on the intestines of rabbits so a company can make health claims about their tea bags – it is deceptive to argue that the experimental animal is used only in life-and-death situations.

"The reality is that an enormous industry has grown around the experimental animal. The infamous Draize eye irritation test—"

"But that's been banned!" Unable to control himself, a speaker on the other team is half out of his seat. The Chair calls for quiet and, as penalty, adds an extra minute to the final speaker's time.

"Yes," she continues, "testing for cosmetics is now banned in Australia, but the manufacturers of many of the products and ingredients we import are free to conduct such testing elsewhere. The Draize test, as I was saying, had been used largely by the cosmetic and skincare industry. Thousands of rabbits were put in mechanical headlocks so staff could pipette a new product in one eye and leave the other eye to serve as a control. Kept in these restraints for two weeks, the rabbits' infected eyes, often scabbing and deteriorating, possibly burning, would be monitored."

She looks up from her cards, and ad-libs. "Imagine, your eye is burning in its socket and you can't touch it." The audience flinches; a few are even scrunching up their eyes at the thought of it.

"Testing on animals for pharmaceuticals is even more problematic, and it is arguable that many toxic tests on animals are done for legal reasons rather than scientific—"

"One minute," says the Chair. Flustered, the girl drops the card she is reading from. Picking it up, she accidentally mixes up the order of her remaining cards. At a loss, she gives up trying to decipher them. She can feel herself turning red.

"What I'm trying to say – sorry, what *we are* saying—" she corrects herself, and turns to face the "for" team "—is that you are not being honest with us.

"You are asking us to trust you, to trust that you know what is and what is not good for us, and to trust what is done to other animals is in our best interests.

"Animals are 'banged' on the head to induce head traumas to study the effects of road accidents, given urinary tract infections and used to try out new contact lenses. They have their spines cut and limbs paralysed to study neuroplasticity and rehabilitation, are forced to binge-drink to understand alcoholism, isolated to study social anxiety, and we have come to the conclusion—"

Finish with a statement, finish with a statement, the final speaker imagines her team silently willing. Their debating teacher, now standing in the back of the room, had pulled her aside earlier. "Throw the punch," the teacher had urged for the hundredth time. "You've got to stop finishing with a question. I think you're scared. Promise me, this time you'll throw the final punch."

*

"Yes, there are alternatives to using animal models," the surgeon concedes. "But not for xenotransplantation, there are no alternatives. Transplants can't be done in vitro because you have to look at the survival of the graft. You need the whole animal."

In one study, he used over 400 rabbits, removing and transplanting their kidneys. This could not be achieved in the isolation of a test tube.

"Xenotransplantation will revolutionise medicine," says the surgeon, his eyes lighting up. "The health of humans around the world would be improved dramatically."

I pause, not in judgment, but because I don't know how to continue. Finally I say, "But to the detriment of another species, of course?"

"Well, yes. Yes, to the detriment of pigs."

"I mean, this would have to be factory-intensive, right? You wouldn't be able to use by-products from pigs already slaughtered?"

"No, of course not," the surgeon rushes to reply. "The pigs would have to be kept in an absolutely sterile environment."

"So they would have to be completely isolated from each other?"

There's a beat. And the opaqueness of the surgeon cracks. He lays his palms, facing upwards, on the table.

"Look, I wish these problems didn't exist in the first place. But they do exist."

Is it ethical to breed animals for spare parts? Is it naive of the surgeon to think that xenotransplantation will revolutionise medicine, that the new skills he has contributed to creating will automatically benefit humans in need of new organs? Or is it cynical of me to think that these achievements, as exciting and groundbreaking as they may be, will simply be used to generate a new market, products and patents that only the developed world will be able to afford? Not to mention creating a whole new factory-intensive industry to burden an already constricting global food source? Or is it all of the above?

And finally, is there really no alternative to xenotransplantation?

What about *us*? Surely we would be the perfect donors for one another? Instead of genetically manipulating a pig, why not take science to its logical conclusion and consider the most precise model of all. Us. Or is it just too unethical to make human organ donorship compulsory?

"This," the surgeon points out, "can get murky. China sources most of its organs from death row." Okay. So because of "transplant tourism" in China, where last year a seventeen-year-old teenager made the news after his parents discovered he had sold his kidney to an online broker for an iPad, because of this we're back to animals. Surely we could at least have a better marketing campaign for organ donation before we turn to animals?

But still, it's messy.

What if the family wants to be near a loved one who has ticked "yes" to organ donation, the warmth ebbing out of their body while surgeons pace, counting the seconds before organs are rendered useless? An organ "farm" would be much more convenient. Plus there wouldn't need to be the frustrating three hours' notice for patients on organ waiting lists to get to hospital for a speedy transfer. And such organs would not be second-hand; they could come with their own guarantee.

Right. So again, what about us? Would it be possible to grow organs from human stem cells? And then clone them?

Well, now. That *would* present an ethical dilemma.

It seems, when it comes to discussing the value of animal testing, no matter how neatly presented, there is always a loose thread, and it only takes the tiniest of tugs to see that this thread leads back to us.

*

So what happened to the third speaker of the "against" team? I guess I lost patience with her.

She probably would have gone on to mention how some manufacturers of carcinogenic products have unashamedly hidden behind unreliable animal test results. Then, in response to the "for" team's description of the miraculous awakening of diabetes sufferers from comas after physicians injected them with insulin drawn from the pancreases of pigs and cattle, she would have countered with the gory details of an Ivan Pavlov experiment in which he rerouted a dog's throat so that its food, once chewed and swallowed, fell through a hole in its throat and landed back into its food bowl. Tethered to a post, the dog kept eating and re-eating this recycled food, while its stomach rumbled for the food that never reached it.

For and against has reached a kind of stalemate.

Scientists today need to be brave. Like their predecessors, those physicians who risked their reputations going against the wishes of the church and the assumptions of their time, and whose achievements we now rely on, today's scientists have their own assumptions to query and commercial hooks to resist.

The irony is that science, with its defence that it is on humanity's side, seeking to ease our suffering, will now have to come up against the pro-life, anti-abortion movement, whose fundamentalists are known to be every bit as rabid as some animal-rights activists. For a human embryonic stem cell to be isolated, a human embryo must be destroyed, and in the eyes of some this is the equivalent of destroying human life. As a result, global research using human embryonic stem cells has faced intense lobbying and suffered considerable obstruction as politicians have played their cards accordingly.

In a national survey in 2009, the researchers Mariah Evans and Jonathan Kelley surveyed nearly 2500 Americans on their attitudes to human embryonic stem-cell research. This is one of the questions in their survey:

Scientists can also grow stem cells without cloning:

BEGIN with a fertilized egg, usually a spare embryo from an IVF program.

GROW the embryo in the lab for a few days until it is a little ball of cells.

DESTROY it and remove the stem cells growing inside it.

GROW the new cells in the laboratory, making lots of the type needed.

TRANSPLANT the new cells into the patient, which may cure them.

UNFORTUNATELY the patient's body sometimes rejects these cells because they have someone else's genes. If that happens, the patient is not cured.

Do you approve using stem cells made this way ...

To replace cells destroyed in a heart attack, if this would certainly save the patient?

To replace cancerous cells, for a person dying of blood cancer?

To cure severe dust and pollen allergies?

Using cloned cells to create new skin, to restore someone's youthful appearance?

Respondents could answer "Definitely yes," "Yes," "Undecided, mixed feelings," "No" or "Definitely not" to each of the options.

The question makes me sad. No national survey has ever asked such a question on behalf of laboratory animals. If they had, to make it realistic, they would have to include hundreds of extra options, plus a simple "all of the above" for respondents to tick.

In spite of efforts around the world to "Reduce, Replace, and Refine" the use of animals in research, it seems that animals – particularly purpose-bred animals – have become a permanent fixture in laboratories, many of which are not exclusively devoted to medical research.

Sure, every now and then there's an "Is this okay?" hoopla in the media, and yes, varying standards of animal welfare are expected to be maintained, while reputable institutions ensure that they have ethics committees to oversee research done in their name – but none of this has seen a decrease in, let alone an end to, the use of animals in research. Some suggest that this is testament to the importance of animal testing for scientific progress, while others suggest that old habits are hard to break.

Like Sultan, the majority of scientists must think the right thoughts in order to follow a certain type of thought process to the end. Do ethical considerations hinder this process? A great many scientists would think so: they are necessary to a degree, but an obstacle nevertheless. But what if it's the bananas – by which I mean the interlocking system of rationality and reward – that are the true obstacle?

Men arrive in clean, shiny 4WDs and utes decked out with kangaroo spotlights and bull bars. They turn into the carpark where the laminated sandwich board says "Bounty" and stop at a desk where two Department of Primary Industries staff wait. From large blue eskies or black garbage bags the fox scalps are tipped into plastic boxes and counted. One man, who shoots for farmers out around the Golden Plains, has helpfully put his scalps in bundles of twenty. "121," the staff call out, taking down the man's details – that's $1210.

After they have been processed, depot staff sprinkle glitter over the scalps to ensure they're not nicked and handed in twice.

Last year, Victoria's Liberal government announced a $4 million "Fox and Wild Dog Bounty." It was the first statewide bounty since 2003. In the years since, the previous Labor ministry ran "Fox Lotto," in which members of shooting clubs could compete for TV sets and motorbikes. Today's scheme, an election promise, is to last four years, with twenty-one temporary collection depots set up in rural areas, eight of which will also accept the scalps of wild dogs. Shooters will receive $10 for each fox scalp they hand in (some shooters prefer to call it a "mask," as it must include the ears and the skin surrounding the eyes and nose). For wild dogs, shooters must bring in a strip of skin that runs from the snout along the back to the tail. Each wild dog will earn shooters $50.

Driving around the back blocks of the eastern highlands, you might come across a tree flanking a paddock, decorated like a Christmas tree with dead dogs tied up and slowly spinning by their hind legs. The bounty gash from snout to tail has left a pink and putrid skid mark, filled with flies. Without a DNA test, it is difficult to tell which of these dead animals are the state's threatened dingoes – although from looking at them, Lyn Watson of the Dingo Discovery Sanctuary and Research Centre reckons at least half could be.

*

"Give them the flat of your palm," says Lyn. "Let them know you through your sweat." From the moment of my arrival, car tyres on gravel, the dingoes have been aware of my presence, their ears pricked up. Standing on benches, boxes, rocks and small hills, they clocked me. Rescued from all over the country, their coats vary according to their home climate: some are golden, others are white or sandy, and there is the odd one in black and tan, "rainforest dingoes," Lyn tells me, "and Victorian dingoes too." Alpine dingoes have tails like feather-dusters "to curl up and keep their nose warm," and the desert ones have tails like rope.

Now, as I try to meet them through the wire fence, most shyly lope away. A few linger near the fence. On Lyn's nod – "This one is friendly" – I tentatively put the back of my hand up to a dingo's snout.

She corrects my greeting. "Give them the flat of your hand. Let them know you. That's right. See, she'll give you her neck soon."

Sure enough, after sniffing and licking my palm, the golden dingo turns her head and presses her neck against the fence. I push my fingers into her fur and scratch under her ear. "She's from a fauna park," explains Lyn. "She's used to people." The others watch from a safe distance.

*

In 2001, a former sheep farmer, Ron Stockwell, successfully sued the State of Victoria for failing to manage the wild-dog population in the state forest adjoining his farm near Corryong, the animals having being classified under legislation as vermin.

"Our paddocks served as a restaurant next door for the dogs," Stockwell told reporters. On ABC's Radio National, he continued:

> Well, you'd go out in the mornings and just go out to do a routine check on your sheep, lambing or whatever, and you'd see … scattered wool and you'd see sheep with blood hanging off them, and then you'd see their guts hanging out on the ground and dead sheep, ewes laying down up against the fence and tied up in the fence, been

chased into fences, and sheep with holes eaten in the guts and you could see unborn lambs inside their mothers, and it's a shocking situation, it would drive anybody off their head to see it.

The case forced the state to step up. Doggers set traps on public and state land, including national parks, and farmers looked after their side of the fence. In 2004, nearly 1500 wild dogs were caught and killed. Old-timers passed on their tricks of the trade to new doggers, telling them to rub eucalyptus leaves on their hands and shoes so they smelt like the bush, to cover traps in leaves and put a stick in front of the trap, so that the dog would step over and into the steel jaws.

Then in 2008, an independent scientific advisory committee warned that the dingo was close to extinction in Victoria. Although they had once been widespread, it was estimated that only one hundred pure dingoes remained. Conservationists intensified their lobbying and the Victorian government acquiesced, listing the dingo as a "threatened" native species. Farmers were furious. It was the first time the dingo had ever been officially considered anything but a pest. But they needn't have worried – the listing was, and is, effectively meaningless. Although protected on remote areas of crown land, dingoes on private property or within a specified radius of such land can be killed. As for shooters in state forests, many still believe that the best dingo is a dead one and there is close to zero policing, let alone education, to change this mindset.

*

So far, on the morning of my visit, ninety wild dogs have been killed and brought in for the bounty. In a later phone conversation with the wild-dog biosecurity manager at the Department of Primary Industries, I ask if the dead dogs will be examined to find out if they're dingoes.

"I don't think so."

"Won't you use DNA testing?"

The manager explains that DNA testing is complicated. It's time-consuming and can give only a probability of whether the wild dog is a pure or dingo hybrid. I catch on eventually that it doesn't matter anyway. If the dingoes are in the wrong place, well, they're in the wrong place.

"They're not protected if they're killing sheep."

"What if they're just walking through?"

"Well. They're not just walking through."

"So there's no program teaching shooters how to distinguish between dingoes and other wild dogs?"

"Visual signs are a very poor indicator when it comes to telling the difference."

"So you don't know if you're destroying dingoes?"

"We don't – we don't believe the wild dogs are dingoes."

"But how do you know?"

"You can tell. Look, I know I said before that you couldn't tell, but you can tell. They don't look like dingoes. Hybrids, yes. But not dingoes."

"Do you know if there are any pure dingoes left in Victoria?"

"No."

"Doesn't this make your work difficult? Not knowing any numbers or statistics?"

"No, not really. Our main focus is dealing with wild dogs predating on livestock."

You get the drift. Foxes, hunting dogs gone wild, dumped dogs, dingo hybrids and dingoes are one and the same. "If it lives like a dingo, then it is a dingo," I am told by numerous supporters of the dingo, even Lyn Watson (although she will only take a dingo into her sanctuary if a DNA test says it is pure).

In the *Australian* two years ago, the biologist Allen Greer wrote that it is "worth remembering that in intact dingo societies it is the dingoes themselves who decide who is accepted and who is not. In other words, dingoes are probably only accepting of animals most like themselves, that is, ones that will fit in."

*

"The dingo is a creature frozen in time," Professor Chris Johnson, an ecologist and a conservation biologist, tells me. "It's halfway between a wolf and a dog."

There are varying theories on how they came to be in Australia. The most popular is that dingoes, having originated from the Asian grey wolf and travelled by foot throughout Asia, were brought ashore by trading seafarers 5000 to 6000 years ago, as a source of fresh food, protection and perhaps companionship.

Australia's indigenous people, however, may have got quite a surprise upon encountering such an animal. The anthropologist Deborah Bird Rose, author of *Dingo Makes Us Human*, wonders how the first meetings of the two species panned out:

> What would it have been like, 5000 years ago, when Aboriginal people and dingoes first encountered each other? The dogs had already learned to live with people, but here were people who had not yet learned to live with dogs. Did these people recognise the common human–dog language? Did the dingoes teach them?

Is "domestication" the correct word to use to describe the dingo/human relationship? Perhaps. But who domesticated whom? If "domestication" doesn't describe what happened, then how to explain the companionable coexistence that developed? Once the dingoes settled in, they quickly became the mainland's top mammalian predator, alongside the Aborigines.

Within the last few thousand years, the number of predators on the mainland has diminished. Previously the marsupial lion, a cat with teeth thought to be stronger than bolt cutters and a pouch for its young, ruled the land. It is believed to have become extinct around 45,000 years ago, after the first arrival of humans. Then the devil and the thylacine became key predators on the mainland. But the thylacine's time at the top was brief. These shy, cat-like wolves, although similar in size to the dingo, were solo creatures. Dingoes were bolder and not only hunted the same prey as the thylacine but sometimes hunted *alongside* the Aborigines.

The dingo returned the Aborigines' gaze. It was the beginning of "three-dog nights," the attempts to domesticate dingo pups and use them for warmth, hunting and, in times of hardship, food. But the domestication was never complete. The dingo retained its independence, its loose family groups and monogamous pairings. To this day, they howl at dusk and at dawn, even in captivity – "Where are you?" "We are here," "Who are you?" "This is me," they say – communicating, reminding and gathering one another into a chorus over vast distances.

Thousands of years later, when the Europeans arrived, modern domestic dogs took the place of dingoes in many Aboriginal camps, and dingoes were driven away. Their style of coexistence could not compare to the submissiveness of modern dogs.

With the running of livestock and bounties for pelts, the persecution of Australia's apex predator began. "Dingo" became slang for a louse and a coward. In 200-odd years, the wolf-shaped, thylacine-shaped and now, in Tasmania, devil-shaped holes in the environment are beginning to show, and biologists are beginning to ask, what happens to a country without its top predators?

*

The most famous example of what happens to an environment depleted of its apex predator is the fable-like story of Yellowstone National Park in Wyoming, America. Established in 1872, it is thought to be the world's first national park, and its story contains lessons not only for farmers, hunters and pest-control schemes, but also for conservationists.

In 1914, in an effort to protect elk and livestock populations, the US government financed schemes to destroy "wolves, prairie dogs, and other animals injurious to agriculture and animal husbandry" on public lands. By 1926, wolves no longer existed in Yellowstone National Park. The result was catastrophic.

Without the apex predator, the elk grew bold. They no longer avoided areas where they couldn't keep an eye out for predators and began to graze in the woodier, denser parts of the park. With their numbers rapidly rising, the landscape was being, quite literally, devoured. Park managers, fearing that the park would soon be denuded, began to cull the elk themselves. But then, in the sixties, hunters complained that there were too few elk, so again the elk were left to flourish and Yellowstone diminished once more.

With no wolves to mark the territory, coyotes moved in. Despite becoming the leading predator, the coyote failed to fill the role of a true apex predator. Because they were unable to bring down large animals, such as elk, the natural regulation of the park's ecosystem could not be restarted. Instead the coyotes drove the pronghorn antelope to near extinction. Yet the fearless elk, no longer controlled by a predator, grazed voraciously; without their habitat, birds disappeared and riverbanks eroded; beavers became all but extinct, their makeshift dams disintegrating; Yellowstone flooded, its topsoil washing away; fish stocks suffered, and the entire hydrology of the area was undermined. The elk triggered an extreme degradation in the landscape that would ultimately bring about a new control, a lack of food for them. As the elk became more numerous and grazed beyond the park's boundaries, financial compensation was introduced to help crop farmers offset the damage caused by these unchecked herbivores.

The entire process was dubbed the "trophic cascade" effect (trophic meaning nutrition) by the scientists Bill Ripple and Bob Beschta, of Oregon State University. Their study sparked recognition of similar scenarios around the world. Even conservationists dedicated to saving the smallest, most vulnerable species, a dedication which inevitably fosters a distaste for predators, are coming to understand that perhaps the best chance for these vulnerable creatures involves restoration of the apex predator to the ecosystem.

The critical role this keystone predator plays in the environment is not what one might assume. It is not simply hunting and killing. An apex predator is ultimately a presence. Mid-level predators are kept in check and their territory is sufficiently limited, while herbivores regulate their behaviour and breeding.

Fear keeps the entire system living within its means.

Australia has the unenviable record of the worst and fastest rate of mammal extinction in the world. Twenty Australian mammal species have been wiped out in the last 200 years, almost half of all mammal extinctions worldwide. These extinctions have stemmed not only from habitat loss and bounties, but have also taken place in seemingly intact habitats. Cats, foxes, rats.

The effect is known among biologists as the "mesopredator release." In the absence of a larger carnivore, carnivores of an intermediate body size proliferate, and populations of birds and smaller mammals are significantly reduced while larger herbivores become overabundant. In Sub-Saharan Africa, where leopard and lion populations have been decimated, the olive baboon is ridiculously abundant. The trophic cascade has gone so far that children are being taken out of school to help their family protect crops from the marauding baboons. Curiously, attempts by humans to suppress one mesopredator often result in the release of another. In Australia, successful fox eradication programs tend to be rewarded with a rapid increase in numbers of feral cats.

trophic cascade.

In 2007, Chris Johnson, the author of *Australia's Mammal Extinctions: A 50,000 Year History*, published a study showing that the presence of dingoes is the most powerful indicator of survival for ground-dwelling mammals and marsupials across Australia. Johnson believes dingoes protect biodiversity on the mainland, largely through suppressing foxes and feral cats, and that good intentions to protect smaller vulnerable species have at times had disastrous results.

In the Tanami Desert, the loss of the last two remaining populations of the mala, the rufous hare-wallaby, a small hopping creature, is one such scenario. In 1987, as dingoes were common in the area and ate the odd mala, the wildlife commission resolved to eradicate them. It succeeded. Within a fortnight, foxes moved in on the first population and ate the lot.

chris Johnson

⚹ book.

Whether the mesopredator would have scoffed the second colony is unknown, because a wildfire swept through that area and the wild mala is now extinct.

In a study published in the *Journal of Conservation*, Mike Letnic from the University of Sydney found that the endangered dusky hopping mouse fared better wherever dingoes were active. If the key predator was absent, however, foxes moved in and mouse numbers declined. Following the dingo fence line, from the Great Australian Bight in South Australia to Coopers Creek in southwest Queensland, Letnic found evidence on both sides of the barrier that the presence of dingoes is beneficial. "Where dingoes were active I found less kangaroos, more grass, less foxes and more small mammals. Where dingoes were missing I saw more kangaroos, less grass, more foxes and less small mammals."

It seems logical, intuitive even, that removing elements of an ecosystem will create weaknesses. But what is more interesting is that the ecology of an environment can collapse following the removal of a key predator. As in a game of Jenga, taking away a single piece can make the whole structure fall down.

In 1995, wolves were reintroduced to Yellowstone National Park after a seventy-year absence. Fourteen wolves, captured by wildlife officials from different packs in Canada, were released in two instalments. Seventeen more wolves were caught and released into the park in the following year.

The wolves did exactly what wolves do. And not only were the numbers of elk reduced, but their grazing behaviour also changed. They moved back to safer areas and the willow, aspen and cottonwood trees grew back. Beaver populations recovered as streamside vegetation returned. The coyote population was reduced by a third, their territory shifted to rocky inclines and away from flat terrain, which in turn allowed for other small predators and scavengers such as foxes, bald eagles and ravens to regain their foothold.

It seems magical and most onlookers are waiting for a "but," yet the only criticism to date has come from hunters who were doing a substandard job as apex predators in the wolves' absence and are now having their permits reduced.

The story of Yellowstone is not unique. Across North America, similar cascading ecological scenarios have been documented, and reintroductions of wolves have been proposed and sometimes followed through. Wherever the wolves appear, so do bumper stickers reading, "Save 100 Elk, Kill a Wolf." And from there, the human opposition to wolves soars.

In 1998, less than six months after twelve endangered Mexican wolves, including a rare alpha female, were released in the Blue Mountains in Arizona, they were reported missing or found shot. The first to be discovered with bullet holes was the alpha female.

For their part, the Yellowstone wolves initially strengthened and rein-habited a wider territory, with some 1500 roaming the state in 2008. Then, three months after they were delisted as a "threatened" species, nearly 10 per cent of the population was shot.

While every other creature falls into line with the apex predator, even mountain lions and cougars, humans continue to resist.

*

It's not too difficult to imagine where our atavistic hatred of predators comes from. After all, we were once prey and opportune scavengers of other creatures' kills. Hunting was once a respectable gig: it is how humans managed to evolve and adapt and survive for this long – not solely because we switched from a purely herbivorous diet to one including meat (although Meat and Livestock Australia would like you to believe this to be the case), but due to our juggling of the roles of prey, predator and scavenger. Humans had to be intense observers of all living things.

"The hunter must become the thing he hunts," wrote J.A. Baker in *The Peregrine* in 1966, revealing how he used the tactics of a hunter to study peregrine falcons for over a decade. And indeed, the naturalist and the hunter were synonymous for centuries. Without the aid of binoculars and photography, birds were shot to be studied, and the collection of specimens from the wild satisfied curiosity, assisted accumulation of knowledge and fulfilled the desire for trophies.

But a split between the two was inevitable. In 1906, the American president Theodore Roosevelt, an avid hunter, popularised the term "nature faker" after weighing in on a controversy triggered by the respected naturalist John Burroughs, who lambasted the growing literary genre of "nature stories," accusing writers of making up scenes in which animals appeared intelligent enough to learn, cooperate and reason.

It was the beginning of an interesting tussle over who loved "nature" more — those who understood nature as a hunter would, or those who saw themselves as observers. In an open letter to the president, William J. Long, who had been targeted as a sentimentaliser, wrote:

> Who is he [Roosevelt] to write, "I don't believe that some of these nature-writers know the heart of the wild things"? I find after carefully reading two of his big books that every time he gets near the heart of a wild thing he invariably puts a bullet through it.

Roosevelt had set aside nearly 150 million acres of forests, established over fifty game reserves, doubled the number of national parks, and among his numerous legacies was the Grand Canyon. And yet, as Long pointed out in his letter, the president went "into the wilderness with dogs, horses, guides, followers, men servants, reporters, and cameramen." His kills required an audience.

And bizarrely, Roosevelt's loathing of so-called anthropomorphism didn't seem to extend to his own writing. In *The Wilderness Hunter*, he wrote of an elk bull as

> a very unamiable beast, who behaves with brutal ferocity to the weak, and shows abject terror of the strong. According to his powers, he is guilty of rape, robbery and even murder.

So Roosevelt shot him. The president loved the wilderness for how it made him feel. Not only did he feel a part of it, he felt on top of it. Man faced beast and man won. The other type of naturalist loved the wild too. Briefly, they treasured moments of inclusion and even clocked an animal in their sights, but they never shot at it.

I am in two minds about hunting. On the one hand, I feel as if there is no longer such a thing as fair game. There is no Moby Dick. Schools of fish are now detected miles away by shipping vessels using military equipment. In America, big-game hunters hire men in helicopters to guide trophy animals towards their rifles.

On the other hand, I'm not opposed to the shooting of a rabbit to go in the pot or a plonk of the fishing line at the end of the pier.

And what about hunting *vermin*? How do I feel about that?

Bill Emmett, based in the Victorian town of Stanhope, is part of a shooters' mob that call themselves "Dad's Army." They have a regular run of farms they hunt on. "All up, there's about seventeen of us." If hunting during the day, they use mostly Jack Russells to flush the foxes out of their dens. "Jack Russells are small, and often we're sending the dogs into blackberry shrubs, so the little dogs get less cut up."

If the fox is pregnant, Emmett says they open her up to have a look.

"And if she's lactating?" I ask.

"A vixen will usually be close to her den, so we'll find it and let the Jack Russells go in and finish off the cubs. It'll take a couple minutes usually. I think the Jack Russells hate foxes more than we do."

There is something about hunting with hate that makes me uneasy.

When I was at a blockade in Tasmania two years ago writing about the island's forest battles, I sat with an activist who had been rebuked by his fellow activists for befriending a young feral cat.

"But I found a home for it in town," he told me.

An older man quickly cut in on our conversation. "You should have killed it." As he gathered momentum, his voice became increasingly righteous. "They're killing everything. An absolute catastrophe."

The younger activist shrugged and then nodded half-heartedly. I tried to imagine this twenty-something waking up to a black-and-white cat purring around his sleeping bag and snapping its neck. "Yeah, I guess," he said.

Then, looking at me, he jerked his head in the direction of a logging road. "Don't you think it's a little bit rich coming from us?" he asked.

"We are the fox," said Andrew Brennan, an environmental philosopher, on the ABC television documentary *Feral Peril*, an investigation into the threat of foxes in Tasmania. "Fox is the mirror of us. We human beings are the true feral species. We are the most widespread, most devastating feral animals on the planet."

feral humans

In poet John Kinsella's essay "Scapegoats and Feral Cats," he writes about the gradual "decaying" of his sanity while working in the wheatbelt of Western Australia, accompanying his colleagues "aged nineteen through to their mid-twenties" as they "spent their evenings down at the local tip, shooting feral cats and their offspring."

> In the red light of an "outback" sunset, I still see D. jumping up and down on an old car bonnet, driving the cats out into the open, and "blowing them away" with his pump-action shotgun. I see my co-sampler with his high-powered rifle, picking others off as they broke away. These people were military in their operation. It made for good stories at the pub, and was met with approval from all there. Cats were vermin and deserved shooting. Furthermore, they deserved to suffer. Half dead, swung around by their tails and flung into the rubbish piles.

Dingo hunting, for the most part, is just as mindless. With the aid of hunting dogs, be it pit bull mastiffs, staghounds, Great Danes, bull Arabs or Jack Russells, shooters seek out the dingoes, sometimes mimicking their wolf-like howls to bring them closer for an easier shot. One wrote on a dingo shooters forum,

> It was a pitch black night and about 2.15am when I came past the
> foal paddock. I spotted a set of eyes so I quickly turned off my lights
> as well as the spotlight … I idled the ute around the stable and
> pointed it in the direction of the gully which I figured the Dingo
> was using to come and go with good cover. I already had my Tikka
> .222 out the window on my rifle rest with a bullet in the breach. I
> flicked on the switch and centred the spotlights beam on the Dingo
> bitch. As she turned to flee my bullet entered her flank and ended
> up exiting the centre of her chest. As I walked out to her with my
> .22 Brno in my hand, another Dingo could be seen just on the edge
> of the light. I moved out of the light so I could see through the scope
> and took a quick offhand shot. The Dog dropped on the spot and
> I had 2 calf killers which I scalped for the bounty.

Calf killers, terrorising, cunning. "They're nasty critters," one shooter tells
me. When I ask him why, he says, "Well, they've been known to stalk you."

Shooters post photographs of themselves grinning, with thumbs up, as
they hold dead pups aloft by their legs, or as pups lie beside them stuck
in traps, not yet dead and leg chewed in an attempt to get free (no need
to check these things for a few days).

"Good job, mate. He's a healthy-looking brute," says one shooter to
another, who has posted a photo of a dead male dingo lying across a tree
stump as he stands in army fatigues behind it. In another photo, a dead
dingo has been propped up to look as if it is alive, its paws standing to
attention.

Nasty critters, hey?

It's a political no-brainer, hating foxes, rabbits, feral cats and wild dogs.
Even local fashion designers have tried to get away with using rabbit fur
in their clothes, claiming that rabbits "are a pest, so it's okay," but I'm yet
to see thousands of white angora rabbits whittling away at the Australian
landscape. Politicians wax lyrical about being 100 per cent committed
to culling foxes and wild dogs, raking in the rural vote, all the while

ignoring the trade of Chinese starfish and Japanese kelp brought over in foreign cargo ships, which is devastating to local marine life – but the undertaking of flushing out a ship's ballast is too expensive, it appears, for large companies to contemplate. Instead we transfer the pest rhetoric to native herbivores such as kangaroos, which, in the absence of a predator, now appear in "plague" proportions.

Of course, I'm being a smart-arse. A lot of this, the importance of the apex predator for the environment, was not known, at least not officially known, until recently. Hell, 200 years ago we were only just getting our heads around the facts that we're related to chimpanzees and not made of stardust.

But today, what excuse do we have? And what alternatives?

*

dogs regulate dingo behaviour

It was January 2002 and dark out when Ninian and Ann Stewart-Moore turned in for the night on their property in northwest Queensland. They were exhausted. They had spent the day picking over the scattered bodies of their merino sheep, some dead, some still alive with their guts hanging out. Dingoes, they'd said over and over, wild dogs. But this time they said it without anger, without those flecks of vengefulness that get one through times like this. This time they were at a complete loss.

No amount of baiting, shooting, fencing and trapping had deterred wild dogs from decimating some 15 per cent of their flock each year. Many of their rural neighbours were switching from sheep to cattle or selling up.

"We couldn't sleep at night, we were in a constant state of anxiety worrying about our flock," says Ninian. "We decided to try one last thing and then, if that failed, to stop running sheep altogether." That one thing was the use of livestock guardian animals, in this case the Italian Maremma dogs. Guardian animals – mainly dogs but also alpacas, llamas and donkeys – are used to protect flocks of sheep, goats, calves and poultry. Highly independent, the animals are bonded to a flock as if it were their pack and look after it by driving predators away.

Dingos, + Guardian dogs in Queensland

In Europe, guardian dogs protected livestock from predators such as wolves and bears for 2000 years, but their use declined when predators were eradicated from most of western and southern Europe. However, in the past twenty years the guardian dogs have made a comeback as conservation projects restore large predators to the European wilderness.

In the United States, Canada and Australia, the guardian dog is making its debut. "We first bought 24 pups for 20,000 sheep," says Ninian, adding that the initial preparation and training of the Maremmas is no picnic. "You can't just chuck them out there and hope for the best. It's a delicate balance: you can't over-humanise them because then they'll just want to live with you, and you can't under-humanise them because they could go feral and become part of the problem." It wasn't just the dogs they had to train – it was neighbouring farmers too. "We lost the odd Maremma from baiting, or a farmer shot them."

But after two to three years of hard work, the Maremmas paid off. "We could count on one hand the deaths of sheep caused by dingoes in a single year. Ten years on, there's about 80,000 sheep in this region, and most farmers are now running guardian dogs."

In the 1970s, millions of dollars were spent on studies understanding the nature of the dingo with the aid of radio tracking equipment. The research revealed that the presence of livestock in the dingoes' diet was at most 2 per cent, but farmers complained that dingoes sometimes chased livestock for the hell of it, stressing them out so that lambing rates were lowered, abandoning their kills or sometimes simply maiming the livestock, chewing the back of their legs or ears. Dingoes, producers said, killed just for the heck of it. Again, it seemed that the ways of the wild and of farming could not coexist.

But the Maremmas at the Stewart-Moores' property are proving otherwise. "I used to think it was about territory," says Ninian, "but we've had a few tracking and collaring studies done, and it appears that dingoes are still encroaching on to what I used to think was Maremma territory, only they're not attacking. There's no stress, no bitten sheep. It's like there's been some kind of dog communication – the Maremmas have indicated the dingoes can pass through, but only if they behave."

At night, Ninian and his wife can sleep easy. Sometimes they hear the guardians yodelling to one another. During the week, they drive around to leave food at the Maremmas' feeding stations. One or two Maremmas might come over for a pat, but most are uninterested. "We can go weeks without seeing them. It's incredible – these dogs are out there, making decisions on their own."

Acting as a kind of handshake between the wild and the rural, guardian dogs have also brought unexpected benefits. For the Stewart-Moores, not only has wool production increased but the quality of the fleece has improved – perhaps due to the reduced stress of predation among the flock.

In Victoria, over a three-year period on Andrew and Glenda Bowran's Riversdale farm in the north-east, wild dog predation saw the loss of 300 adult sheep and only fifteen of approximately 650 lambs survived to adulthood each year. In 2006, the Bowrans began bonding Maremma guardian dogs to their flock. Currently four Maremmas are guarding the sheep, with no adult sheep lost to predation and lambing up by 70 per cent.

The *Best Practice Manual for the Use of Livestock Guardian Dogs*, put together by researcher Linda Van Bommel for the Invasive Animals Cooperative Research Centre, reveals that the Bowrans' Maremmas sometimes wander over to a neighbour's property and look after their sheep during lambing. As a result their predation losses have also decreased.

And it's not only livestock that guardian dogs are protecting; conservationists also use them to protect wildlife. Today guardian dogs keep foxes away from penguin and gannet populations along the Victorian coast.

*

"The Maremma are a good piece of reconciliation," says Chris Johnson. Now he and his fellow conservation biologists want to bring back the predator. Not just the dingo, but also the Tasmanian devil.

"Whoa, are you serious?" I ask Johnson on the phone.

"Well, yes," he says, mildly. "In the Grampians, we could get the foxes and cats right down and then introduce the devils—"

"For real?" I cut in. "This is going to cause a huge hoo-ha! Wait till Andrew Bolt hears about it!"

"Yes, well, I suppose it will, but in ten to twenty years we hope to see devils back on the mainland, in Victoria at least, where they used to be incredibly common—"

"Oh my god," I shriek, unable to contain myself. "This is a total GAME CHANGER."

Poor Johnson has no idea how to take me. But I love the thought of bringing predators back. Feral cats and foxes may as well put out deck-chairs and drink cocktails, their habitat is so damn relaxed at the moment.

"People don't like the concept because they think it's too simple," Johnson tells me — and it's true, I'm worried I like the concept of reintro-ducing the predator just because it lets other species do the ugly work. But is it really that simple?

Today, few of us in the Western world will experience the fierce first-hand reality of being prey — prey of another species, that is. In 1985, Val Plumwood (formerly Routley), a renowned Australian environmentalist and intellectual, was death-rolled by a crocodile three times. Fifteen years later, she wrote about the encounter in the essay "Being Prey." She wrote of being in a "red hot pincer grip," death-rolling in a "centrifuge of boil-ing blackness," clawing her way up a mud bank, bloodied, mangled and surprisingly alive.

> The wonder of being alive after being held — quite literally — in the jaws of death has never entirely left me. For the first year, the experience of existence as an unexpected blessing cast a golden glow over my life, despite the injuries and the pain. The glow has slowly faded, but some of that new gratitude for life endures, even if I remain unsure whom I should thank. The gift of gratitude came from the searing flash of near-death knowledge, a glimpse

"from the outside" of the alien, incomprehensible world in which the narrative of self has ended.

Plumwood went on to explore the modern position of humans outside the food chain, the idea that animals can be our food but we can never be theirs – even in burial, our bodies are sealed tightly from the soil, resisting the natural process of becoming fertiliser.

> The thought, *This can't be happening to me, I'm a human being, I am more than just food!* was one component of my terminal incredulity. It was a shocking reduction, from a complex human being to a mere piece of meat. Reflection has persuaded me that not just humans but any creature can make the same claim to be more than just food. We are edible, but we are also much more than edible. Respectful, eco-logical eating must recognize both of these things. I was a vegetar-ian at the time of my encounter with the crocodile, and remain one today. This is not because I think predation itself is demonic and impure, but because I object to the reduction of animal lives in factory farming systems that treat them as living meat.

"Large predators like lions and crocodiles present an important test for us," continued Plumwood. "An ecosystem's ability to support large preda-tors is a mark of its ecological integrity. Crocodiles and other creatures that can take human life also present a test of our acceptance of our ecological identity."

Today, in America, some ranchers are now advertising "wolf-friendly" beef. Will Australia one day be selling "dingo-friendly" wool? Do we have it in us to coexist, to restore predators to the wilderness and share the vast range of land that they need, even though some may present a threat to us?

Australia, in a sense, has it lucky. If restoring the top predators to eco-systems is as essential as the evidence is revealing, then at least we don't have to consider the logistics of reintroducing leopards, grizzly bears, tigers and the like.

Here, two human deaths since colonisation are attributed to dingoes. In 2001 nine-year-old Clinton Gage was killed by two dingoes on Fraser Island, and in 1980 it's now widely accepted that Azaria Chamberlain, a two-month-old baby, was taken out of a tent by a dingo at Uluru. But is this two too many? On Fraser Island, where one of the last populations of the pure dingo remains, rangers are culling dingoes that appear over-familiar with or threaten humans, most of whom are tourists swarming to the island every year to see these very same dingoes, some throwing their sandwiches at the creatures in order to get a closer look. Can we modify our behaviour without resorting to tabloid fear campaigns and calls for culling?

*

It comes back to us.

"Look at their ears," says Lyn Watson, pointing at two dingoes playfully chasing one another. Perked and triangle-shaped, the dingo on the chase has made her ears face forwards to fully take in her target, while the dingo being chased has spun his ears around to listen to her approaching footfall.

Watching them from a distance, Lyn points out the dingoes' quirks. As they inspect their burrows ("That den over there, under the log, is huge. We never saw them making it. Who knows what they did with the dirt!"), check one another's "pee mails" and splash in wading pools, it would be easy to conclude that the dingoes here are content.

"But they're not dogs, they're wolves," says Lyn, waiting for the day when she can return her dingoes to the wild.

When we were little, my brothers and I, and all the other kids on the street, stood beneath a gum tree that hung like a leafy chandelier over the back alley and called up to a flock of white-feathered, yellow-mohawked cockatoos.

"Hello cocky, hello cocky, hello cocky," we chimed until one cockatoo, somewhat reluctantly, replied, "Hello." We fell over each other laughing. We knew this cocky, he had been our friend's pet cockatoo before he took off with the mob.

"Hello cocky," we called up again and again until he gave in, quietly, as if shamed in front of his new flock.

"Hello."

"Half the pleasure of having a dog, I could see, was storytelling about the dog," wrote Adam Gopnik in the *New Yorker*, and I can't think of a more accurate statement to describe my relationship with animals.

Since I can recall, encounters with animals, their cameos, curious habits and visits, have formed the basis of stories, fragments of poems, childhood touchstones.

There were the green tree frogs my mum brought home from the fruiterer. They snuggled up with the bananas in Queensland, only to wake in Melbourne being rudely flushed down the drains by grocers. Today, a woman visits grocers around Melbourne with special eskies and arranges for soft-hearted truckies to take the frogs back up north on their next haul.

There was "Superhen," a chick my brother hatched at school that wooed the rest of my family with rocket-shaped eggs while terrorising me on the sly. No doubt with my eldest brother's collusion, Superhen had decided I was below her in the pecking order.

Then there was Tiger, a stray and easygoing tomcat that stole my family's heart, stunning us with an out-of-character lightning-quick paw, a card-playing "Snap" from under the dinner table to make off with a schnitzel.

As a child, I found solace in animals. Some will say I anthropomor-phised them. I gave them roles to play in my life, during the long after-noons I spent with my rabbit, and then cats. Perhaps they were simply furry vessels for me to fill, and I was actually having conversations with myself. This is possible. After all, I was convinced Tiger shared my feel-ings about Superhen. He and I once climbed a tree to escape, only for her to flap up to our branch and keep pecking us. I could have sworn the look on Tiger's face was the same as the look on mine. Horror.

So yes, perhaps I anthropomorphised the animals in my life. Yet I also loved being with them because they were not human. Strangely, even then — *especially then* — as an eight-year-old I had moments when I felt less alone with animals than I did with humans.

But I admit I may be more far gone than most. In the city, I detour past the block of flats on Lygon Street with the creamy porous façade that seems to be a haven for dozens of cocoons.

Once, riding my bike through Carlton Gardens at midnight, I saw a bird, a species that I had first met in Brisbane, a nankeen night heron, an awkwardly hunched bird, cheek feathers the colour of a bruise setting in, and I crunched on my brakes so hard that the bike flipped. When I looked up from my tangle, the heron was still standing there, looking at me ever so sadly.

At an animal sanctuary in rural Victoria, as a volunteer led me around pointing out miniature pet pigs and horses abandoned by their owners after turning out to be not so miniature, a peacock that had been dumped at the gate one morning suddenly took centre-stage, displaying his great fan of tail-feathers, shimmering the eyes and slowly spinning as if at a bar mitzvah.

The nearby pigs and ducks paid no attention. A chook clucked past. Only we two humans stopped to gaze and take in the bird's stunning beauty. What is it about us?

*

"Us" and "them" — what lies between is a pockmarked battlefield. Bomb-shells of upset, tipped-over mythologies, convenient theories and strata of knowledge. The theory of evolution and genetics, theories of automatons, gods of dominion and stewardship, mumblings of prayers, the sharpening of knives and the hissing of stunners, the measuring of brains and craniums, and every now and then the odd release of "them" to come join "us," or vice versa.

In Susan Sontag's *Regarding the Pain of Others*, she wrote about an exhibition in New York that displayed photographs taken in small towns across the United States from the 1890s to the 1930s. "The pictures were taken as souvenirs and made, some of them, into postcards," wrote Sontag; "more than a few show grinning spectators, good churchgoing citizens as most of them had to be, posing for a camera with the backdrop of a naked, charred, mutilated body hanging from a tree."

The postcards were of trophy kills, lynched black men. Sontag wrote of her "obligation to examine" these photos, to

> understand such atrocities not as acts of "barbarians" but as the reflection of a belief system, racism, that by defining one people as less human than another legitimates torture and murder. But maybe they were barbarians. Maybe this is what most barbarians look like. (They look like everybody else.)

As Hannah Arendt noted, acts of atrocity are often committed under a veneer of banality. The great evils in history are not necessarily the work of solo psychopaths, but of ordinary people — who accepted, observed and normalised unspeakable acts against sentient beings.

Animal activists often compare the historical treatment of women, slaves, indigenous people and children as property and inferior beings to the way we treat animals; and indeed, the rationale, the rhetoric, often used to maintain the status quo is unnervingly familiar.

American farmers claimed economic ruin would follow the end of slavery. Public figures claimed that granting rights to women was plainly absurd and, until recent decades, it was widely believed that babies could not feel pain and were operated on without anaesthetic (which was thought potentially dangerous), with only a muscle relaxant administered to stop the baby from thrashing about on the surgery table.

When selecting slaves to ship from Africa, some merchants were said to check the saltiness of their skin as an index of suitability for the high seas; others weighed up the merits of "loose packing," which involved loading fewer people to reduce disease and ensure at least a minimum profit at the journey's end, and "tight packing," which was simply cramming them in, come what may. As with the production of foie gras, a device was invented to pry open their mouths to force-feed those who were attempting suicide by starvation. Is it too much to connect this to what live animal exporters now call "shy-feeding"? On the long journeys, shackled slaves, unable to jump overboard, were recorded as having begged one another for strangulation.

And once in America, runaway slave notices had their own newspaper column, with owners describing their property as having "downward looks" and "lurking" in various areas:

> The New-York Gazette, August 31, 1730.
> Run Away last Tuesday two Negro men, both branded RN on their shoulders; one remarkably scarrified over the forehead, clothed with trousers, The other with a coat and trousers. Whoever brings the said negroes to Jason Vaughn in New-York, shall have thirty shillings and all reasonable charges paid.

In his work *Against Liberation: Putting Animals in Perspective*, Michael Leahy conceded that animals can "manifest relatively short-term distress at, say, the loss of a mate," but emphasised that the distress, the emotional capacity of an animal, is short-lived and limited. In other words, they get over it. "Without language," Leahy stated, "it cannot *consider* its plight."

"I would not hesitate for one moment to separate any half-caste from its Aboriginal mother, no matter how frantic her momentary grief might be at the time. They soon forget their offspring," said James Isdell, whose title was Protector of Aborigines, in 1906. *They soon forget their offspring*. This we now know, and perhaps knew then, is not true.

Of course, there are problems with what I am doing. Humans are not the same as other species. Becoming a free man, free woman or free child involves a handshake of sorts, an acknowledgment that this particular "them" is actually "us" and has the right to join us – be it at the pub, the ballot box, in the workplace, at the bank, in the pews, at home, and so forth.

Other animals cannot become *us*.

Sure, we've created paler versions, washed-out chimp actors who like to watch themselves on television and flick through rodeo magazines, remembering their days as stunt animals. And the old saying that dogs look like their owners, and vice versa, has never been more valid, with numerous humans using their pet as an extension of their personality.

But *they* will never be *us*.

Yet this is not to say that the presumptions, the unspeakable acts, the ideologies, the excuses and the contempt that our species has, time and time again, put into effect against one another and against animals are not bound up together.

Several years ago, David Weisbrot, then president of the Australian Law Reform Commission, announced that "animal protection may potentially be the next great social justice movement." In 2005, the University of New South Wales became the first law school in Australia to teach animal law. Now, ten Australian universities are teaching it, while some high schools offer it as a legal studies elective. Yet, like environmental law, the field is still rather nascent. Lawyers cover a range of animals, from pets to strays, wild to circus animals, livestock to greyhounds, and prosecute cruelty cases on their behalf, manage estates that are left to beloved pets and animal societies, act in custody battles, and so on. Practitioners dust off existing laws that are rarely enforced and argue for legislative reform.

So questions regarding our relationship with, and our treatment of, other animals are going to keep surfacing. If we create an animal, are we free to delete it? If, by our own standards, we slaughter an animal humanely and adequately manage its suffering, is it ours to do with as we will? Is it better to exist to be slaughtered, or not to exist at all? How to ensure that the butcher, the scientist, the farmer recognise that the creature in their care is a being, even as all the while they continue to use it as an object?

> "I believe it's completely feasible,'
> said Bob Rust
> of Iowa State University,
> "to specifically design
> an animal for hamburger."

So goes a poem by John Berger. It's difficult to forget the image of bobby calves being thrown like packages in the post, hurried along like unruly parcels, and I wonder, *What's next?* Will we try to breed out their bellows? Their expressions of pain? Or better yet, how about we breed out their sentience? An animal that wants to die, that runs to the knife. What then?

No matter how reluctant we are to face these questions, there are more and more people out there determined to ensure that even if we are unable to answer them, we at least acknowledge their existence. Many of these discussions will inevitably turn in on themselves, the importance of animals becoming a discussion of why they're important *to us.*

For example, there are numerous studies which reveal that cruelty inflicted on animals can lead perpetrators to inflict similar cruelties on their fellow humans. A recent advertising campaign by the RSPCA drew on this, featuring the bruised face of a child with a stamp reading, "Tested on Animals." I am, however, loath to quote from such studies for fear another study emerges showing that beating a dog is a better outlet for one's anger than, say, beating a girlfriend. This possible inversion, like so much of the linguistic gymnastics when it comes to discussing the "value" of animals, makes me wary of making specific claims about why animals are important.

So, I will not argue that maintaining ecosystems, a planetary balance, is essential to the survival of our species. In my book *Into the Woods*, I wrote that many biologists are beginning to describe the modern geological era as the Anthropocene, the sixth in a series of extinctions, all said to be caused by extreme phenomena, in this instance by the harmful activities of man. "Of course," I wrote,

> this cannot be proven for certain, at least not until it is well and truly upon us, a risk that many seem willing to take. But perhaps more poignant is biologist Edward O. Wilson's description of the period that will follow. Wilson says it will be "the Age of Loneliness" – a planet with us and not much else. In his writing there is no apocalypse, no doom, no gates of hell, no wrath of god, or mass hysteria – there is only sadness.

But even here I'm using the notion of *us* to highlight the importance of *them*. I'm seeing the world's species as a long string of light bulbs: if we keep letting them go out, we'll be left forlorn under a single bulb. Is there no way I can convey the importance of animals without falling back on relative conceptions?

*

In the *Mackinnon Project* newsletter, a publication on livestock health and management, Andrew Fisher, a professor at the University of Melbourne's veterinary school, warned livestock producers that by attacking animal welfare and animal-rights groups instead of tackling the issues at hand, they were using up their remaining political capital and residual public goodwill.

> Australia, with its foundation stories of the outback, clearing and settlement of farms, and sheep and cattle grazing, still produces a strong resonance of life on the land in many people's minds, even

Never again miss an issue. Subscribe and save.

☐ **1 year subscription** (4 issues) only $49 (incl. GST). Subscriptions outside Australia $79. All prices include postage and handling.

☐ **2 year subscription** (8 issues) only $95 (incl. GST). Subscriptions outside Australia $155. All prices include postage and handling.

☐ Tick here to commence subscription with the current issue.

PAYMENT DETAILS Enclose a cheque/money order made out to Schwartz Media Pty Ltd. Or please debit my credit card (MasterCard, Visa or Amex accepted).

CARD NO. ☐☐☐☐ ☐☐☐☐ ☐☐☐☐ ☐☐☐☐

EXPIRY DATE   /   AMOUNT $

CARDHOLDER'S NAME

SIGNATURE

NAME

ADDRESS

EMAIL   PHONE

tel: (03) 9486 0288 **fax:** (03) 9486 0244 **email:** subscribe@blackincbooks.com **www.quarterlyessay.com**

An inspired gift. Subscribe a friend.

☐ **1 year subscription** (4 issues) only $49 (incl. GST). Subscriptions outside Australia $79. All prices include postage and handling.

☐ **2 year subscription** (8 issues) only $95 (incl. GST). Subscriptions outside Australia $155. All prices include postage and handling.

☐ Tick here to commence subscription with the current issue.

PAYMENT DETAILS Enclose a cheque/money order made out to Schwartz Media Pty Ltd. Or please debit my credit card (MasterCard, Visa or Amex accepted).

CARD NO. ☐☐☐☐ ☐☐☐☐ ☐☐☐☐ ☐☐☐☐

EXPIRY DATE   /   AMOUNT $

CARDHOLDER'S NAME   SIGNATURE

ADDRESS

EMAIL   PHONE

RECIPIENT'S NAME

RECIPIENT'S ADDRESS

tel: (03) 9486 0288 **fax:** (03) 9486 0244 **email:** subscribe@blackincbooks.com **www.quarterlyessay.com**

Delivery Address:
37 LANGRIDGE St
COLLINGWOOD VIC 3066

No stamp required
if posted in Australia

Quarterly Essay
Reply Paid 79448
COLLINGWOOD VIC 3066

if they are uncertain of modern farming realities. It would be to our farming industries' ultimate detriment if this residual goodwill were reduced through mismatched expectations on issues such as environmental management or animal welfare.

To my mind, these foundation stories have worn awfully thin.

The American writer Michael Pollan, author of *The Omnivore's Dilemma*, once said that, "Food should be alive, and that means it should eventually die." The problem with factory-farmed meat isn't the meat, believes Pollan, it's the factory. This is a view that has more supporters by the day. Even Mark Zuckerberg, creator of Facebook, announced last year that he would only eat what he himself killed.

Of course, this could all backfire. The US television hit *Portlandia*, a comedy that takes the piss out of the granny-frocked, thick-rimmed-glasses-wearing, fixed-gear-bicycle-riding hipster enclave of Portland, Oregon, has its characters not only wondering about the ethical origins of the chicken on a restaurant menu, but also about the origins of the chicken feed. Ethics is, indeed, a slippery slope.

Slippery in so many ways – as J.M. Coetzee has written, the Nazi concentration camps were modelled on the Chicago slaughterhouses.

What are you trying to say? I can feel my editor urging. *Don't finish with a question. Throw the punch.*

What I want to say is, let's not kid ourselves. The injustice is complete. This is not a debate over whether our treatment of animals is unethical or not. It's unethical. We know this.

"That's nature," say some.

"Survival of the fittest," say others, mangling the theory of evolution.

"There's no other way," a couple pleads.

The question is: just how much injustice do we want to partake in?

Interestingly, in each case study I've written about in this essay, I've encountered a softening, an understanding of the vulnerability and fragility of other creatures, as a person faces their own mortality.

In the chemist and writer Gail Bell's essay "In the Rat Room," she writes of her own research work involving animals, noting that as she has gotten older, a sense of unease and doubt has crept in surrounding the use of animals for scientific research.

> With age comes reckoning. And more, if we are to believe findings published in a recent edition of *Psychology and Aging*. "Emotional intelligence" peaks as humans move into their sixties. Older brains appreciate the "sadness" of sad situations because the "detached appraisal" switch that was employed in youth breaks down. This is interpreted as an evolutionary uncoupling of pathways and network systems that were built up in the storms of shaping a life out of the materials to hand and that are now largely redundant. Without battles to fight, the "brain on ice" (Lenin's words) begins to thaw out.

The surgeon admitted to me that he couldn't even bear to fish anymore, while in Ralph H. Lutts's *The Nature Fakers: Wildlife, Science and Sentiment*, he reveals the hunter and president Theodore Roosevelt's growing uncertainty in his old age. In a letter to the naturalist Burroughs, Roosevelt writes, "I must say that I care less and less for mere 'collecting' as I grow older."

> A week later he wrote to Burroughs again to tell him how accustomed the animals seemed to be to the presence of humans at his holiday home.
> "Really I have begun to feel a little like a nature faker myself," he confessed, as he wrote of the events of the past two weeks. Describing how a chipmunk would cross the tennis court during a match, he quipped, "I suppose that Mr Long would describe him as joining the game."

There is no greater brutality than that found in us: our dumb-headed hatred of pigeons even though it was we who invited them to live with us, they who learnt to obey the flick of our wrists, gestures on a rooftop, and to deliver messages for us. Our mob-like hatred of pit bulls without any thought for what the breed and its assorted individuals have been through at our hands. In 2007, the American quarterback Michael Vick was found to be running an illegal dog-fighting operation. In the woods near his home, police discovered two car axles buried in the ground where pit bulls were tied up and used as "bait" for other dogs. Those that showed no taste for blood were used as bait and, if not killed, were shot or hooked up to batteries and electrocuted. Dogs that showed fighting potential were injected with steroids and made to hang from bars by their jaws to strengthen their grip. Over fifty-one pit bulls were seized that day, many of them petrified of humans and each other, covered in scars and cigarette burns.

And, bizarrely, there is no greater love for *them* than our love. One couple in Tasmania told me how a wombat they had rehabilitated and returned to the bush found his way back to their home a few years later – sick and walking with a limp. After they had restored him to health, he returned to the bush of his own accord.

Animals inspire us. Over and over humans drew them in caves and on rocks. We studied them, we made tools and we spoke in order to get closer to them. The idea of film, the moving image, came from studying a bird's flight, a flick-book of frozen images suddenly coming to life. Would we have ever thought of flying if it weren't for birds? Swimming if it weren't for fish? Submarines if it weren't for whales? They inspire us to buy a new mobile-phone plan. To their detriment, animals inspire us. To their advantage, animals inspire us.

Free of charge, they inspire us.

"I came late to the love of birds. For years I saw them only as a tremor at the edge of vision," wrote Baker in *The Peregrine* – and I wonder, just how fragile do we need to become to understand that animals are important?

That is all.

Peter Hay

How daunting it is. How scientists and public actors of enlightenment and principle – journalists, politicians, bureaucrats, commentators, activists – can be immersed in the ichor of climate change politics without surrendering to futile desperation is beyond me. I can't – the enormity of what is happening and the utter inconsequentiality of our puny collective response disarms me entirely.

So it was as I pondered the statistical nightmare relentlessly laid out by Andrew Charlton in his *Quarterly Essay*. And yet Charlton is able to look the looming catastrophe squarely in the face and remain optimistic. Ah, but he's an economist, of course, and he has his ready solvent at hand – when trouble looms large, human ingenuity, prompted by market signals, goes to work and technological genius repairs all, growth spirals away, and progress, the conceptual cornerstone upon which Western civilisation rests, resumes its axiomatic trajectory onwards and upwards into eternity.

I will come back to this. But let me acknowledge the strengths of Andrew Charlton's essay. They lodge within his frank, often brutal, description of the problem. Climate change is the central challenge of our times, and its resolution is of incalculable importance. Charlton vigorously argues that Australia's emissions reduction target of 5 per cent below 2000 levels by 2020 is not a piece of gutless tokenism, but extremely ambitious, and probably unattainable.

What nettled, though, was the reduction to cardboard cut-out cliché of so much complexity and nuance in the many and various opposing positions. It is much easier, for example, to dispose of environmentalist arguments against GM food if you simply ignore the most compelling element in the case against it – the very real danger of rogue genetic escapees contaminating other lifeforms, with uncontrollable and disastrous consequences.

Similar objections can be made to Charlton's depiction of issues of global equity. It is disingenuous in the extreme to characterise this as a conflict between selfish First-World environmentalists careless of the legitimate needs of the world's poor, and those seeking to emerge from poverty into prosperity. Leaving aside the observation that there is little point in developing countries increasing their consumption of energy and their material throughput if the end result is the destruction of the conditions that make human life possible — that global eco-logical wellbeing must logically constitute a necessary pre-condition within which the goal of the good life is to be pursued — it is simply not the case that green prescriptions ignore the interests of the 6 billion who live in poverty. The evidence that this is so abounds in the relevant literature. Here, for example, is Clive Hamilton in *Requiem for a Species*, urging, at page 223, "vigorous political engagement" that does not "abandon the poor and vulnerable to their fate while those who are able to buy their way out of the crisis do so for as long as they can."

I'm looking back over the rich factual underlay upon which Charlton builds the case for his "Plan B," his strategy for beating a path through the formidable "Kaya Identity." Wonderfully well argued, but there's an element missing, and without it I don't think we can make it. It's this: the 1 billion living on "islands of prosperity in an ocean of poverty" need to reduce significantly their own levels of material consumption, because it is not possible, even, I think, with "Plan B" new technologies, for the entire world to consume at the levels that pertain in the rich countries. Then, and only then, can we make the moral case that the path taken by us is not one that others should seek to tread, that it was a mistake from which we are now in retreat, and that there are other ways of being prosperous, and of measuring prosperity, in the learning of which we in the West will need to be cast as students. Charlton does not argue for this. Again, this is something to which I'll briefly return.

Charlton overlooks the environmental activism that is entrenched in many developing countries, an activism that, among other things, defends traditional agricultural practices over industrial models of agriculture imported from the West via the loan policies of the IMF and the World Bank. Vandana Shiva and Arundhati Roy are two who have cogently argued that Western development models have destroyed local agricultural sustainability (by requiring farmers to grow cash crops for export, and by closing off rights to collect and save seed locally); that women in particular, traditionally the custodians of local agricul-tural practices, have been disempowered by aid programs; that capitalisation of agriculture has destroyed the social and economic fabric of local communities, fuelling destitution and a massive shift of population into the shanties of the burgeoning cities; and that the major recipients of grant and loan moneys from the West (including the relevant UN agencies) are the new and fabulously wealthy entrepreneurial classes in the major cities. Charlton looks at none of this. It falls below the broad sweep of his gaze. But, given that this is all in keeping with his one-dimensional, conventionally Western notion of "development," he presumably approves. I do not.

But it is a good essay. Remember that as I construct my catalogue of dissent. It is a good essay. Its major flaw is that to which I alluded at the beginning of my commentary – the author's unbridled technocratic trust. I know, so far as the discipline of economics is concerned, that such an article of faith constitutes the party line. But it is dangerous. Because faith is what it is – a Hail Mary that, cometh the hour of crisis, cometh the men and women of genius to supply the technological breakthrough that will set civilisation back on a steady keel. It is not an empirical truth, not a law of history – it is essentially a prejudice, almost religious in character. Charlton argues that it has ever been thus – that the advent of crisis has invariably seen the technological breakthrough necessary to restore equilibrium and rescue civilisation. As an antidote to such dangerous nonsense we should note that virtually all civilisation-threatening crises were technologically induced in the first place, just as the climate change crisis is today. And we should note that history is actually a litter of civilisations that have flowered and fallen precisely because they did not solve the crises with which they were beset – usually crises of resource depletion, often accompanied by shifts in ambient environmental conditions. Sound familiar? Even Western civilisation – the one predicated upon the blithe assumption of never-ending progression from a given state today to a better state tomorrow – even Western civilisation does not conform to the Charlton theory of history. The Dark Ages constituted a major rupture, for instance, from which a starting over needed, eventually, to emerge. And one of these days there will be another crisis where the magic of the technological solvent will fail.

Charlton is himself curiously inconsistent in his technocratic trust. He places his faith in new technologies rather than more efficient technological applications of known technologies, arguing that the necessary investment is more likely to cohere around the former than the latter. But this is a bifurcation that is difficult to sustain. Given the uncertainties involved in moving from technological breakthrough to economic feasibility, it seems more likely that technological investment, given appropriate government policies, will seek to discover more efficient ways to deliver existing technologies – and this includes the renewables, the targets for Charlton's selectively applied technological pessimism. Improvements in renewable energy technologies to the point that they might begin to meet base-load power demands seems at least as feasible as the rapid development of some of the currently nowhere-in-sight technological options listed as promising by Charlton.

The cobbler, they used to say, should stick to his last. I don't entirely agree with this – polymaths are welcome on my turf, and Charlton has a capacity for wide-ranging synthesis rarely encountered in an economist. Nevertheless, his "optimists" versus "romantics" dualism is downright silly. For a start, those environmentalists so caricatured are not romantics. Romanticism was a movement without an ecological sensibility. It was intensely individualistic. Nature was not valued for itself, but as a spiritual device – an instrument through which the refined individual sensibility could, via contemplative engagement, attain heightened enlightenment. Such a "nature study" aesthetics of the individual is not the ecologically derived sensibility of the contemporary environmentalist. The romantics were nostalgic, looking back to a golden past from which humankind had unaccountably strayed. Today's environmentalist has almost no interest in the past, tending to regard it as a series of option-narrowing mistakes. And the romantics derived, from their intense individualism and their yearning-back, conservative political and social philosophies. Those whom Charlton would label "romantic" are at the forefront of political and social progressiveness.

Then there are the optimists, those who believe in a triumphalist vector of history, in "the power of human progress." But an extraordinary thing has happened. Such a coalition should have included, as its very standard-bearers, the Republican Right of the US and the Australian Right within the Coalition, for these are the believers in material progress *par excellence*. But climate change science has driven a wedge between the optimists, such that it no longer makes any sense to deploy this label in the way Charlton does. Suddenly science is not the handmaiden of unremitting progress, but is found to point emphatically in the opposite direction. Faced with such an appalling eventuality, the fundamentalists of the unconstrained market economy have embarked upon the greatest mass demonstration of cognitive dissonance in modern times. They have chosen the sacred truths of ideology over the clear conclusions of science. It is a development that has thrown the very authority of science as an enterprise into crisis – and that in itself is a development of civilisation-threatening consequence.

The climate change crisis has created a similar legitimacy crisis for the democratic polity. It requires of people a level of technical sophistication that most do not have, thereby fostering reversions to foolish but tenaciously embraced prejudices. Despair suffuses all. Anger has become the currency of what passes for democratic exchange, and political rhetoric is ramped up to levels of provocation that democratic life can't sustain, while neither political leaders nor the crusaders within the commentariat seem aware of the extent to which the tone and terms of political discourse – which they largely determine – are destroying democratic legitimacy. For these developments the climate change debate is largely, though not exclusively, responsible.

And I am one of the many who have fallen off the truck. I have been an ultra-democrat all my life – I still describe myself as a "strong democrat." But the television news comes on and I leave the room. Because the public realm has become morally and intellectually threadbare. Faced with the crisis of the millennium the democratic estate shows itself to be powerless, puny, pathetic. Why did Andrew Charlton omit that necessary component from his "Plan B" – the requirement for the affluent West to opt for drastic cuts to its own level of material opulence? Here's the answer on page 44 of his essay: "In rich countries, policies to cut emissions that come at a large cost to households simply will not succeed." And he's right. We will not step back from unsustainable opulence. So I despair, my prognosis for my planet hopeless. For all our sakes, let us hope that Andrew Charlton is right, and I am wrong.

Peter Hay

Eric Knight

In the opening pages of *Man-Made World*, Charlton sets us up for a titanic clash. He paints the scene at Copenhagen in December 2009 beautifully and writes that it exposes the central dilemma of our century: choosing between economic progress and environmental sustainability. However, as the essay continues, Charlton reveals that this is a false choice. We must find the balance between the two, he tells us. In so doing we must rely on something very powerful: technological innovation.

In making the case for technology, Charlton is bang on the money. He deserves credit for making it so articulately. If anything, he is at times too coy about the faith we should place in innovation. Charlton sometimes gives the impression that technological innovation is mere luck. Von Humboldt's discovery of fertiliser in the late nineteenth century was a "lucky event" and a "chance discovery." When stockpiles were exhausted, Fritz Haber's invention of a synthetic substitute emerged when "providence smiled." Charlton is right to acknowledge the revolutionary nature of these technologies, but to put them down to luck alone is to confine the hope of human progress to a gamble. This is not necessarily so. Luck plays a role, but equally (if not more) important is good management.

Luck is a popular theme in how we narrate our stories of success. We are, after all, the Lucky Country. Our mining boom is fortuitous. For technology sceptics, a technical solution to climate change is a fluke. Too rarely are we compelled by something often closer to the truth: we can make our own luck.

When it comes to the climate, to acknowledge the importance of technology strongly influences where we focus our attention. It moves us away from condemning consumption and focuses the mind on the driver of technological innovation: the market. The market is not always well calibrated to invest in creative ideas, and we have seen distortions in recent years that have rewarded alchemy over entrepreneurship. But when we get it right, engineers and entrepreneurs – guided by the clockwork of competition and reward – can steer us towards answers we never thought were possible.

This, to be clear, is not a do-nothing approach. It is effort conscientiously directed towards a solution other than de-industrialising the world and capping consumption. The policies required to harness market forces to develop clean-energy technologies deserve detailed attention elsewhere, but the broader point – that the challenge is more than merely cutting consumption – is an important one. This conclusion will undoubtedly raise the ire of some on the left, but it must be defended on humanitarian grounds. As Charlton correctly points out, a non-technological solution stalls the aspirations of 6 billion to ascend out of poverty. We must let people be free to pursue cars and canapés if they so desire, but we must strive to make these carbon-free.

Where Charlton's analysis begins to drift is on what it would take to build a political consensus around these ideas. His analysis of the left is deep, but his analysis of the right is thin. "The right remains immutably sceptical in the face of mounting scientific evidence," he writes. And he returns to a familiar framing of our political choices: "Kevin Rudd ... called climate change the 'greatest moral challenge of our time,' while the Opposition leader, Tony Abbott, described the science as 'absolute crap.'" Such a frame dooms the quest for political consensus to failure. It is like setting a multiple-choice test where none of the options offers the answer.

Genuine political consensus on this issue will be formed only by building a meaningful bridge to the right. For that to start, the difficult issue of scepticism needs to be unwrapped. My view is that the "believer versus sceptic" choice is a false one. Too few of us have the qualifications and experience to make a reasoned judgment either way on the science. The best we can do is to choose whom to trust. This has always been a question of delegated authority rather than science. In a world where we lack the technical skills to make certain judgments, whom should we trust?

Scepticism, then, is really rooted in a question about the kind of democracy we want to live in. Are we comfortable in delegating authority to expert scientists, and, if so, what is the scope of that authority? It is one thing to delegate to scientists on matters of scientific opinion. It is quite another to take their views on the economic policies we should apply in response. The boundaries of authority around climate change have long been blurred. Is Al Gore an authority on science or politics? What about Nigel Lawson? Is James Hansen, the head of the NASA Goddard Institute of Space Studies, a scientist or a policy adviser? In 2008 he publicly offered advice to President Obama on "solving the climate and energy problems, while stimulating the economy." The same can be asked of Bjorn Lomborg: scientist or economist?

Acknowledging that this debate is really about *authority* rather than science, *democracy* rather than morality, is crucial. We need to refresh our thinking before we can build political consensus. The alternative – the impulse to banish any mention of scepticism from this debate – leads to absurd outcomes. In 2009 Raj Pachauri addressed an audience in Abu Dhabi on climate change. At the time Pachauri was the head of the Intergovernmental Panel on Climate Change. There was no room for doubt on the science, he argued. Elsewhere he likened sceptics to the modern-day equivalent of the Flat Earth Society. Yet doubt is the very premise of scientific enterprise. *Nullius in verba* – take nobody's word for it – is the motto of the Royal Society. It propelled the scientific revolution through the sixteenth and seventeenth centuries. Science is filled with doubt, but we must learn to make judgments that favour the balance of probabilities at any particular moment in time.

What, then, to do? In searching for a political consensus, Charlton makes one of the boldest statements in the essay. "You might think the answer is to provide people with more information, so that consensus can emerge from greater understanding. But you'd be wrong." Charlton moves in the wrong direction here. It is true that more information per se does not solve problems. But if we are to keep faith with our democratic institutions, we must continue to disclose information freely and widely, trusting in people's reason to find the right balance on issues.

Our greatest challenge is not that people are irrational or ignorant. Rather, it is that they are often asked to answer the wrong questions. A community that must choose between progress and planet will be divided. It is a false choice. Likewise a political community that is told to choose between "absolute crap" and our "greatest moral challenge" will be deeply confused, even offended.

This is not the real political choice we must make. Our politics has become used to simplifying messages in pursuit of complex reform. I think this trend pulls in the wrong direction. Instead, we must trust people with more information, not less. And we must think carefully through the question of delegated responsibility. A healthy democracy delegates some responsibilities to experts, but not too many. The art is in finding the right balance. We should trust our ability to resolve this kind of complexity and resist the temptation to manage it on behalf of others from the top down. If we fail to do this, we will lose much more than just the Great Barrier Reef.

Eric Knight

John Quiggin

As an economist working on climate change and trying to communicate with interested members of the public, one of the greatest challenges I've faced is to explain the magnitude of the likely costs of meeting Australia's climate targets for 2020, and of decarbonising the economy in the longer term. In particular, I've talked about the estimated cost of a program that does enough to limit warming to around 2 degrees Celsius (relative to pre-industrial climates). Such a program will require Australia and other developed countries to cut emissions by around 90 per cent by 2050.

Most such estimates suggest that if a market-based approach were adopted globally, income per person would be about 1 to 5 per cent lower by 2050 than if no further action were taken. Since Australia is an energy exporter, the impact here would be higher than elsewhere, so I'll use the upper end of the range and suppose that a comprehensive mitigation program would reduce our income by 5 per cent by the time it was fully implemented.

Is that a lot, or a little? Australia's national income (this is a more relevant measure than the more widely quoted GDP, but the two don't differ much) is currently around $1.2 trillion a year, so a reduction of 5 per cent would be around $60 billion a year. Over forty years, using standard discounting procedures, the aggregate reduction in income could be around $2 trillion.

That sounds like an awful lot, but in reality it just illustrates our difficulty in handling large economic numbers over long periods. To put the same estimate differently, the best estimate of our likely average growth in income per person in the long run is around 2 per cent per year, which implies that income per person will double in about thirty-five years. An ambitious mitigation program would reduce the rate of growth by around 0.1 percentage points, and delay the time at which income would double by around two and a half years.

An important implication is that, in terms of overall living standards, the typical household will barely notice the impact of such a change. Household incomes are quite volatile, commonly varying from one year to the next by 10 per cent or more. That's more than twice as much as the entire impact of a climate change program over forty years. Even for the aggregate economy, the difference between a recession and a boom is more than 5 per cent. Obviously, effects of this kind vary a lot from household to household.

Another useful comparison is with health care, which currently accounts for 9 per cent of GDP in Australia. On current projections of demographic changes and developments in medical technology, it's very likely that this share will increase to 15 per cent by 2050 (it's already around this level in the US). That change would be similar, in economic terms, to the impact of decarbonisation.

Changes of this kind require a substantial reorientation of economic activity, with some industries contracting substantially and others expanding. But there is nothing special about this. Over the course of the twentieth century, employment in agriculture fell from 30 per cent of the workforce to 3 per cent. After peaking at almost 50 per cent of total employment in the middle of the century, the combined share of manufacturing and mining contracted to around 20 per cent by 2000. The recent mining boom has not reversed this trend, as modest growth in mining sector employment has been offset by continued contraction in manufacturing.

Provided the economy is growing, adjustments of this kind usually go fairly smoothly. Something like 20 per cent of Australian workers change their jobs every year, so the labour market can certainly accommodate structural changes far larger than those that will be associated with decarbonisation. Still, the impacts will not be trivial, particularly at the regional level. Some parts of the country will do better and others worse.

If the costs for Australia are significant but manageable, how would a global agreement to decarbonise the economy affect the world as a whole? It depends on the kind of bargain that is struck, of course, but the most obvious basis for an agreement is "contract and converge." That is, all countries should eventually converge on a common entitlement for emissions per person. Countries with higher emissions would buy tradeable permits from those with lower emissions.

For an agreement on this basis, Australia's costs would, as I've already suggested, be at the high end. The cost for rapidly developing countries like China and India would be less than 5 per cent of national income (at current rates, around six months' worth of growth). There is no fundamental conflict between a sustainable climate and continued economic progress.

Why has it been so hard for me to explain this to non-economists? The debate over energy seems to have led people to think in binary terms. Either the problem of climate change is one that can easily be solved with a few small adjustments (turning off light switches, some modest subsidies for renewable energy and so on) or it requires changes that amount to the end of industrial civilisation as we know it. There is, it seems, no middle ground.

This brings me, at last, to Andrew Charlton's essay. His title implicitly endorses the binary choice of options I've criticised above. In the end, however, he argues that there is a way to resolve this dilemma, namely that we should look once more to technology and innovation for solutions.

Focusing on the Australian debate, Charlton correctly observes that the adjustments required even to meet the government's 2020 target of a 5 per cent reduction in emissions will be too large to be achieved with the currently proposed carbon tax. If the target is to be met, we will have to resort to importing emissions credits, a path that can't be pursued forever and obviously is not available to the world as a whole. As Charlton says, "Pretending that addressing the problem will be without cost is a recipe for failure."

But, as is typical in this debate, Charlton then jumps to the opposite extreme. He claims that price-based policies like carbon taxes or emission trading schemes cannot possibly achieve the necessary reductions in emissions, and that it is necessary to pursue an alternative path, namely that of large investments in research to produce a technological solution.

There are two big problems here. The first is that there is a large element of magical thinking in the invocation of a technological solution. Charlton gives us no reason to suppose that the research program he advocates would be more cost-effective than the investments (including in research) that would result from the adoption of a carbon price. (So if a carbon price would cost too much, in terms of foregone growth opportunities, to be feasible, the same would probably be true of a publicly funded research program.)

But the much bigger problem is that, having shown that a carbon price so low as to be effectively cost-free will not achieve the necessary reductions, Charlton offers no serious analysis of whether the goal can be achieved at costs that are significant, but still affordable. He makes vague references to billions, but in an economy that generates over a thousand billion a year in output, that is scarcely conclusive.

The most substantive analysis Charlton offers is in his discussion of Australia's target for 2020, which he addresses using the Kaya Identity. Charlton states that "energy efficiency can be assumed for this analysis to continue to improve at its historic rate of about 1 per cent per year." That assumption is reasonable if the price of carbon is so low as to have only modest effects on demand.

But suppose that the price were raised to $100 per tonne as the Greens have suggested. That would add 10 cents per kilowatt-hour to the cost of electricity generated by burning black coal. At that price, the energy used by an inefficient fridge or air-conditioner would be substantially more than its purchase price, producing a strong incentive to seek more efficient alternatives, or to focus on less energy-intensive forms of consumption. The incentives for business to reduce energy use would be even sharper. It seems reasonable to suppose, under such circumstances, that the annual rate of improvement in energy efficiency might rise to 2.5 per cent and that the growth rate of demand for energy-intensive forms of consumption might fall below the rate of GDP growth, say to 1 per cent.

Redoing Charlton's calculation with these assumptions radically changes the conclusions. Now the demand effects of higher prices would reduce energy use by more than 20 per cent over ten years. On that basis, only 10 per cent of existing emissions would need to be replaced. Moreover, the price incentives associated with a $100 per tonne carbon price would make both wind and solar renewable energy competitive with gas. Existing brown coal plants would become uneconomic, as would new investments in black coal plants.

Charlton implicitly concedes this point, saying, "Most of the scalable clean options are so expensive that the carbon price required to encourage genuinely commercial investment would need to be in the range of $100–500 per tonne, not the $20–30 now planned." These numbers are exaggerated (most estimates suggest that wind power needs a carbon price of $50 per tonne, while recent cost reductions make solar PV competitive at less than $100). More importantly, Charlton does not even try to make a case that a price of $100 per tonne is not feasible.

So would a carbon price of $100 per tonne, phased in over ten years or so, be economically ruinous? It's easy to see that it would not be. The government's current policy is expected to raise about $10 billion in revenue, most of which will be redistributed back to households. If the price were four times as high, and produced a 25 per cent additional reduction in emissions, the total revenue would be around $30 billion after the contraction in the tax base was taken into account. That's less than the revenue raised by the GST.

To put things more bluntly, the objections to a carbon price high enough to stabilise the global climate are political, not economic. The problem is not that a price-based climate policy would destroy economic growth but that it would create (in fact, has already created), in Charlton's words, "a political firestorm."

Advocacy of a technological solution, which is never going to be funded on the scale necessary to achieve the desired outcomes, amounts to a choice to back away from a problem that, for the moment, appears to be in the political "too hard" basket.

John Quiggin

Ben McNeil

In January 1984, the late Apple co-founder Steve Jobs was excited to showcase the Apple Macintosh, the world's first personal computer with a version of the graphical user interface and mouse we use today. Displayed onstage during its launch, the Macintosh spoke disparagingly of its IBM rivals: "Hello, I am Macintosh. Never trust a computer you cannot lift." Despite this, not everyone *could* easily lift it, as it weighed a hefty eight kilograms.

Today, some twenty-seven years after that launch, I frequently notice my two-year-old twins throwing a small black 100-gram object back and forth like a toy. That little black object is called an Apple iPhone and it is so easily tossed around by my kids that I don't know how it survives the ongoing barrage of violence. Despite its near weightlessness, it holds 150,000 times the data of the original Apple Mac, is hundreds of times more powerful and is not just a computer; it's a life-absorbing revolution.

Now capable of creating photos and video, communicating and tweeting, the computer is much changed. The original Macintosh, a heavy slab of hardware that could barely process one of today's email messages, also cost some $5400 in today's dollars, an enormous expense. The benefits of computers were clear, but only the wealthy could afford them. In 1984, someone could have easily shown calculations and facts on the enormous cost, weight and area needed to process growing data demands using those clunky, slow, expensive computers. In fact, someone could have called the area of land needed "Computistan." But those people would have been oblivious to the future, not through malign intent, but because they were captured by the constraints of their time, incapable of seeing how technology would change in just a couple of decades.

Technological progress comes with something else, too: growing prosperity. Only a few decades ago, Taiwan and South Korea were among the poorest nations in the world. At the beginning of the personal computer revolution, those countries took a long-term bet, investing heavily not only in research and development, but also in business subsidies and support for innovation around micro-processing, computers and, more recently, smart-phones. They saw the long-term potential in "Computistan" and today are prosperous nations because of it. The expensive, slow, clunky development of the computer in California, and its slower spread into poorer nations, didn't hinder economic progress – it boosted it, nearly everywhere.

Andrew Charlton is an intelligent, articulate and welcome voice in Australia's public-policy debate, and he makes some important arguments in his *Quarterly Essay*. The essay nicely details the immense past and future environmental pressures on our world, and Charlton quite rightly shows how traditional environmentalists have misjudged the power of technology in overcoming population and food pressures. But Charlton makes just as big a misjudgment on clean technology because of one trait that plagues most political advisers and economists: short-term thinking. I can't help wondering whether if Charlton were advising Prime Minister Bob Hawke in 1984, he would have obsessed about the massive cost and technical difficulties of "Computistan," or advised of the enormous possibilities for Australia and the world in the longer term. After reading his essay, I can't help but conclude that he would have been stuck thinking about typewriters when he should have been thinking about the coming revolution.

When Charlton calls green growth (or the notion of economic growth arising from environmental gains) "nonsense," he is technically correct. Energy costs in Australia will go up when a price is put on carbon pollution, which will slightly reduce economic growth rates in the short term. The same logic, however, was true when compulsory superannuation was introduced in the 1990s or when trade tariffs were lowered in the 1980s. There are short-term sacrifices, but in the long term the wider benefits far outweigh the initial costs, and this is where his argument comes unstuck.

Charlton uses a wonderful quote from a former Saudi oil minister – "The stone age did not end for lack of stone" – yet he somehow misses its green-growth context. The businesses and nations who developed the "post-stone" solutions were the ones that prospered economically, as will the nations that develop the "post-carbon" solutions of tomorrow. Clean, low-pollution technologies are a long-term competitive advantage in a world that will inevitably crave cleaner cars, trains, planes, houses, offices, materials, food, water and, I almost forgot, energy: this is what I have called the Clean Industrial Revolution.

Green growth is why China has invested more in recent years than the US in clean technology, and why President Barack Obama declared in his 2010 state of the union address, "The nation that leads the clean energy economy will be the nation that leads the global economy." Marius Kloppers, the CEO of BHP, said in late 2010 that "Australia needs to look beyond just coal" and price carbon if it is to remain economically competitive. Kloppers wasn't doing this because of any moral imperative: he sees there is a race on to find the technologies that will help solve these global challenges.

To see an example of "green growth" (or to see the long-term competitive advantage in low-pollution technologies) we need look no further than the global car market. In the 1990s, oil was dirt cheap at US$20 a barrel; petrol was US$1 a gallon in the US. American carmakers like General Motors and Ford developed larger and larger vehicles since the American consumer didn't care about fuel economy when petrol was so cheap. However, with rising populations, dwindling oil reserves, Middle East tensions and the cost of carbon emissions from oil use, the price for oil was always going to go up in the longer term. Yet American carmakers, instead of investing for the long term (as well as the short), simply denied that fuel efficiency would be important to their future. By 2007 (while the economy was still booming) Ford reported a loss of US$1.4 billion and General Motors a staggering US$39 billion. Meanwhile, Toyota reported a profit of US$17 billion and Honda $5 billion – both rising on the back of booming sales of their more fuel-efficient vehicles. Developing lower pollution technologies meant *more* profit, not less. This is green growth.

Progress and the planet are not mutually exclusive and for Australia's recent senior economic adviser to miss the wider green-growth economic story is disappointing, but symptomatic of the political debate we have in Australia, particularly on climate change.

Charlton also argues that a 5 per cent cut in emissions for Australia by 2020 is "ambitious" using the Kaya Identity model. That model assumes historical "business-as-usual" emissions growth, and because Australia has historically been the most carbon-obese developed nation in the world, this model would *always* tell us that a 5 per cent cut in emissions is ambitious. Let's go back to 1984 again.

While Steve Jobs was launching the Macintosh in January 1984, Australia's best economists, experts and modellers were busily working away in the departments of Treasury and Prime Minister. As they do every year, they were working on economic growth forecasts. To do this, they have to assume future levels of productivity and the structure of Australia's economy. The problem is, economic models can only project the future based on the past; they can't predict how technologies or external political forces will make the future "business-as-unusual."

So those economic modellers in Canberra in the early 1980s issued a twenty-year forecast to the prime minister that would have excluded not only the information technology sector and the hundreds of thousands of jobs in it by the year 2000, but also the enormous productivity and growth in our wider economy that has come from the use of those technologies. And like those economic advisers to Bob Hawke in 1984, Andrew Charlton, in calling a 5 per cent cut "ambitious," is assuming what happens in the future is based on what happened in the past. In fact, even before Kevin Rudd became prime minister, things were changing. By 2008, 97 per cent of Australia's *new* planned energy investments were either in renewables or natural gas, while Malcolm Turnbull's banning of inefficient incandescent light bulbs in 2007 and the widespread uptake of solar panels and insulation in homes and businesses has created a new lower carbon pathway for Australia's economy. These structural changes make Charlton's assumptions (and so too Treasury's) obsolete.

It is partly because Australia has such high levels of pollution per unit of economic production that it is much easier to cut emissions. In the 1980s, the United Kingdom had a coal addiction like Australia's. But that made it much *easier* to reduce emissions rapidly. The UK *cut* total emissions by over 15 per cent between 1990 and 2002 (it was actually 10 per cent within eight years) while increasing its population by 2 million and expanding its economy at a similar rate to Australia. How? It wasn't renewables or even nuclear power that did it, but switching from coal to natural gas. It is because Australia is so coal-addicted today that switching to cleaner energy options and boosting energy efficiency gives much greater marginal savings. Charlton's essay is fixated on options like wind, solar, nuclear and clean coal. Yet one of the biggest opportunities for Australia to lower carbon emissions quickly while maintaining energy supply is a shift from coal to natural gas. In particular, natural gas tri-generation involves small-scale power plants that use both heat and waste efficiently for other purposes, driving down emissions by 90 per cent in some cases and having economic productivity dividends over the longer term. This is how the city of Sydney is planning on hugely cutting emissions over the next ten years.

One final point: carbon pricing versus greater spending on research and development is a chicken-and-egg debate. If pricing mechanisms are "not particularly effective in bringing forward the technologies of the future," then how was it that the 1970s oil-price shocks were the beginning of private and public investment in solar power research, wind power in Denmark and more advanced nuclear power reactors in France and the US? Carbon pricing will work to encourage private investment in clean technologies; it just depends on the price levels and certainty for business. Pharmaceutical companies invest billions in developing new drugs because they know our health will always be valued. Carbon pricing for the long term will help boost those private sector clean-tech start-ups. The problem lies in the politics. The federal Opposition has vowed to repeal the carbon price if elected. The current political environment couldn't be worse for encouraging private investment in clean-tech start-ups and research.

For the developing nations, Charlton's argument is correct. The billion people living on $2 a day need more energy in the quickest and cheapest way possible. And my experience at UN climate negotiations has been just as disheartening as Charlton's. But this hasn't made me pessimistic; on the contrary, seeing the immense new investment from nations such as China, Brazil and South Korea has made me more optimistic. A good thing about economic globalisation is that when a technology is developed, like the computer, it is spread much more readily than in the past. China today is rolling out underground fibre-optic cables, not the archaic above-ground telephone wires that dominate the Western landscape. Leap-frogging old Western technology is progress for developing nations. For water, food and houses, the spread of cleaner technologies is what gives me greater optimism that we can achieve both "planet and progress."

Ben McNeil

Charles Berger

It sure is refreshing to see an economist of Andrew Charlton's stature who doesn't believe markets will fix everything. His basic premise that "there is no choice between progress and planet," that we must reconcile economic and environmental goals, is surely correct.

And few environmentalists would dispute his main practical conclusion that Australia should complement a price on carbon with "huge direct government support for new energy infrastructure and a massive increase in research funding." Indeed, that's pretty much what most of us have been saying for a long time.

Yet Charlton's unrelenting sniping at "green groups" is puzzling, and a major blemish on his otherwise mostly cogent analysis. According to Charlton, "green groups" are guilty of starving the world by supporting organic agriculture and opposing genetically engineered organisms; criticising Australia's emissions reduction target as being too weak when really it is rather ambitious; opposing nuclear and large-scale hydroelectric power; "glossing over" the difficulty of shifting to renewable energy; talking a lot of innumerate "nonsense" about green jobs; and, worst of all, ignoring global poverty by promoting a carbon price against the interests of poor countries.

I'll take up each of these charges, but first a question. Why does Charlton single out "green groups" for intense, specific and sometimes personal criticism while remaining silent or vague about the positions and records of those with the real power to implement the policies he advocates?

Where is his criticism of the economic rationalists who have taken over swathes of our public service, the fearful and myopic politicians who have gutted the very research programs he thinks should be accelerated, the rent-seeking businesses who routinely hold governments to ransom and paralyse any sensible policy process, the self-serving commercial press that gives any old climate-denialist wingnut more airtime than a Nobel laureate? There is barely a whisper about them. This all strikes me as akin to criticising the firefighter for holding the hose a bit wrong, while letting the arsonist off scot-free.

Let's get down to some specifics. Charlton rejects organic agricultural techniques on the basis that "organic farms produce lower yields per acre, even as the world needs more food with less environmental impact." This is simplistic and partly wrong. A recent CSIRO summary of relevant research is more nuanced: "yields equivalent to or better than conventional agriculture may be achieved, although often they are not; … yields decrease during conversion but then improve afterwards."

And yields are not all that matters: increasing food *security* is more important in most places than increasing raw yields. Hunger can and does occur amid ample food supplies. Resilience and dependability are crucial, especially for poor farmers. Industrial monocropping may increase overall yield in a good year, but leave a poor farmer far more vulnerable to disease, disaster, market fluctuations and exploitative conduct than a more diversified crop.

In drought-prone areas of Australia, new techniques like intercropping and pasture cropping can result in slightly lower average yields, but much more dependable yields in years of poor rainfall. Ask a farmer in the Wimmera which he prefers. And Cuba's experience in the 1990s following the collapse of the Soviet Union shows that societies are able to feed themselves using much less petroleum and nitrogen fertiliser than is usual in conventional agriculture.

Charlton's conclusion that organic farming doesn't work for the poor doesn't sit comfortably with the fact there are 50,000 organic farms in Mexico alone. These exist not because poor Mexican farmers have been duped by green groups, but because these farmers have determined that organic farming actually works well for them, providing a more diversified and resilient crop than industrial techniques. And little wonder, since organic farming makes most sense precisely in places where labour is relatively cheap and farmers don't have the cash for expensive imported fertilisers, fuels and pesticides.

Organic farming isn't a panacea; nothing ever is. But by dismissing it out of hand, Charlton engages in exactly the kind of knee-jerk, one-sided argumentation he accuses green groups of.

Conversely, Charlton is critical of green groups that oppose genetically modified food. Forgive our scepticism. We're still trying to get over the rock-solid assurances we were given that cane toads would stay in cane fields.

On "clean coal," Charlton advises green groups to drop their opposition, even though he concedes the trials have been "extremely disappointing." And so here we are accused of being too bullish on one technology (renewable energy), which is "not yet well developed enough to perform at sufficient scale and reliability," but we're told to back another technology (carbon geosequestration) that is at an even earlier and more disappointing stage of development. Geosequestration may work in a few special cases (such as oil and gas extraction sites), but nobody in the energy industry thinks we're anywhere close to economically viable application at coal-fired power plants.

Charlton takes issue with green groups, specifically the Australian Conservation Foundation, that criticised the 5 per cent pollution reduction target set by the Rudd government. He seeks to make 5 per cent an "ambitious" target by a bit of quantitative chicanery. Charlton assumes population growth, economic growth and energy efficiency are all fixed and asks us to consider how challenging it would be to meet a 5 per cent target by changing our mix of electricity generation alone. But population, economic growth and efficiency are not fixed; they are all variables over which Australia has some control. His assumption that energy efficiency will only improve at 1 per cent per year is far too pessimistic; it assumes we can do no better at reducing energy demand than we have done historically. Yet many of the largest and most cost-effective strategies for reducing emissions are precisely in this area. In effect, Charlton asks us to imagine addressing climate change with no new strategies for improving public transport, no electric cars, no major improvements in the efficiency of buildings or industrial processes, no improvements to interactive household and business smart meters, no biological sequestration, and no shift in our lifestyle and consumption patterns. His focus on supply-side solutions ignores any possibility that we could achieve far more ambitious targets through a wider range of demand reduction strategies.

Charlton's conclusion that "we must accept that attempts to bring about a legally binding global agreement ... have failed" appears premature in light of the modest but definite progress achieved at the Cancun and Durban climate talks. There's a big difference between having failed and having not yet succeeded. Progress is nowhere near as fast as most sensible people would like, and the later a global agreement is pushed out, the harder the task becomes. But not a single nation has walked away from the table so far. We greenies are nothing if not bloody-minded and we plan to keep at it for as long as it takes to get a global agreement, even if Andrew Charlton has lost interest.

His claim that "carbon pricing doesn't work for the poor" is wrong. No sensible green group has ever suggested that Burkina Faso pay the same price for a tonne of pollution as Australia, or indeed that Burkina Faso should pay anything at all. Most proposals allow for some form of "contraction and convergence," whereby poor countries are allowed to increase emissions, while rich countries cut theirs. The primary consequence of carbon pricing for very poor countries is to unlock significant new revenue streams for carbon sequestration projects through REDD (Reduced Emissions through Deforestation and Degradation) and similar mechanisms. Properly designed, a global carbon price would see poor countries as direct and significant financial beneficiaries.

Charlton takes us to task for "glossing over" the difficulty of shifting to a clean-energy future through an emphasis on renewable energy. Here he may have a point, yet in a sense his faith in technological solutions suffers from the same flaw. By implying that our environmental and resource challenges can all be solved by more investment in infrastructure and research, his essay avoids any confrontation of deeper political, cultural and social practices. And while I agree with the need for such dramatically increased investment as part of the solution, Charlton doesn't address any of the barriers standing in the way. Will we increase taxes or slash other spending to fund this investment? How can such a huge shift be prosecuted quickly through our current political system? Charlton's insights, given his history in Canberra, would have been useful here.

Notwithstanding all of the above, I can only endorse in the strongest terms Charlton's core recommendation for "huge direct government support for new energy infrastructure and a massive increase in research funding." Yes, a thousand times over. It is for this reason the Australian Conservation Foundation puts immense effort into advocating for policies like the Clean Energy Finance Corporation, which will direct $10 billion into clean technologies in coming years. And green groups have consistently gone in to bat for basic science and research capacity. Technological innovation must be a core component of our efforts to create a better future. There is no inconsistency between seeking to supercharge our innovative and creative spirit and at the same time questioning some aspects of an overly consumption-oriented society.

Charlton obviously has a lot more insight into ecological questions than the rigid economic rationalists who hold sway within his profession. I wish there were more like him. And I think there's far more common ground with green groups than his essay suggests. Andrew, let's talk.

Charles Berger

Jeremy Baskin

Andrew Charlton is one of an emerging and influential cadre of environmental policy "realists" (Stewart Brand and Mark Lynas are others) who both recognise the gravity of our current environmental situation and whose prescriptions are often made in the name of tackling poverty and enhancing development. But like many politicians in developing countries, these realists who look both level-headed and fair-minded may also be charged with a version of "hiding behind the poor."

This is not a charge of bad faith. These thinkers are aware of the depth of the planet's environmental problems and the current limitations of renewable energy technologies. They are both concerned with global justice and realistic about the obstacles to reaching any agreement on action. But three elements of their position are troubling. First, they tend to blame environmentalists (largely seen as Western and rich) for promoting unworkable approaches to "planet" that ignore the importance of "progress"; they are rather less critical about the role of more powerful economic forces in society. Secondly, they are unwilling (or too "realistic") to challenge received wisdom about poverty, which traditionally sees "progress," "development" and growth as the answer. Finally, and perfectly logically within this paradigm, they gravitate towards technology as the answer (including extreme technophilia such as geoengineering).

Charlton's essay is a telling instance of this orientation. He argues that for developing countries to "progress" and emerge from poverty they need "more, and cheaper, energy." Climate change will not be solved "by reducing energy use or making dirty fossil-fuel power more expensive."

To illustrate his arguments, Charlton makes much of the construction of South Africa's massive 4.8-gigawatt Medupi coal-fired power plant. For Charlton, this major investment is essential for post-apartheid South Africa to bring electricity to the millions not yet connected to the grid and to assist them to emerge from poverty. He implies that opponents of this US$13 billion World Bank–supported project are environmentalists, comprising both rich-world governments and NGOs who place "planet" ahead of "progress" (thereby ignoring the poor?).

A more nuanced account of the Medupi project may undermine Charlton's arguments. South Africa is already a middle-income, high-emissions economy (with higher per capita emissions than many rich countries), where inequality is extreme and poverty and unemployment are widespread. Further, South Africa has suffered major power outages in recent years and demand management by the state utility Eskom has proved insufficient.

The expansion of supply through Medupi is aimed at providing major industry with electricity – including under long-term, low-price contracts, made in the dying days of apartheid, for projects such as BHP Billiton's Mozal aluminium operation in neighbouring Mozambique. It is not, to any significant extent, aimed at linking ordinary poor, rural and semi-urban households to the grid. Indeed, doing so would probably be achieved more cheaply, resiliently and effectively by decentralised generation at a smaller scale.

Medupi could be seen as essential to tackling poverty only by relying on assumptions about the knock-on effects of supplying more energy to big industrial and mining operations, leading to more economic growth. Unfortunately, since the end of apartheid in 1994, while the economy has grown by approximately two-thirds, unemployment has risen (and today stands officially at around 25 per cent, but in reality is more like 40 per cent), as has inequality. This suggests that traditional grow-the-economy approaches to poverty are not working. It may also suggest that the root problem is inequality, and poverty is the effect.

Charlton is correct in stating that more households have been connected to the grid since apartheid ended. Unfortunately connections are one thing and electricity flows through those connections are another matter. Those on the ground will tell you that the major trend has been rising disconnections, mainly prompted by poor households being unable to feed their pay-as-you-go meters.

Opposition to the Medupi project has hardly been restricted to the well-heeled Western environmental lobby, as Charlton implies. While such opponents have received greater publicity, in large measure because of their literacy levels and media savvy, there has been a raft of community complaints, many of which the World Bank complaints panel has conceded are well founded. These include concerns about health impacts, the strain on water resources in an area vulnerable to extreme shortages, the impact on subsistence and commercial farming in the region and on the substantial eco-tourism industry.

More generally the project leaves South Africa — a country which is highly vulnerable to climate change — with infrastructure that will lock the country into a high emissions, high water-use trajectory for decades to come. It also carries the dual risks of reinforcing the cheap electricity expectations of major resource companies, as well as foreign currency-denominated risk exposure through the substantial loans taken out to fund it.

It is true that there is no neat answer to the real demand in developing countries for more energy. And there are cost and scale issues associated with renewables (although the Medupi project is particularly expensive per megawatt-hour). Hence the conclusion Charlton reaches may turn out to be unavoidable for some of the very poorest, smallest low-emissions economies. But in general, locking developing countries into high emissions options is no solution at all, not for the planet, nor for those particular countries, nor even for "progress." At best it is a subsidy of this generation by the next one. And of the rich by the poor in the name of "development."

When Charlton argues that "environmental challenges are a distant threat compared to the daily tragedies of life in slums and villages," he is correct only in the most superficial sense. He ignores the multi-directional connection between poverty, inequality and environmental destruction. And he is simply wrong — or at least his progress/planet binary has come back to bite him — when he says those who are focused on environmental challenges are usually based in rich countries. He assumes that what Roderick Nash called "full-stomach environmentalism" is all or most of environmentalism, when arguably "empty-belly environmentalism" is more widespread.

I can do no better than end by quoting Ramachandra Guha, writing about the work of Indian journalist and environmentalist Anil Agarwal. First, that "the main source of environmental destruction in the world is the demand for natural resources generated by the consumption of the rich (whether they are rich nations or rich individuals and groups within nations)." Secondly, that "it is the poor who are affected the most by environmental destruction" and that the "eradication of poverty in a country like India is simply not possible without the rational management of our environment and that, conversely, environmental destruction will only intensify poverty." In short, we need to spend at least as much time rethinking "progress" as rethinking "planet."

Jeremy Baskin

Robert Merkel

Necessity may well be the mother of invention, but it is not a sufficient condition for it to occur. The alchemists spent centuries searching for the Philosopher's Stone, and the desire to turn lead into gold is as strong in 2012 as it was in the Middle Ages. But, in 2012, do we obtain our precious metals by transmutation from cheap ones? Hardly. It remains far cheaper to dig gold out of mines of a scale and depth unimaginable to the alchemists than to perform neutron transplants on atoms of mercury.

As such, I note with some unease the central contention of Andrew Charlton's essay: that for developing nations to take action on climate change, it must not impede in any way their progress out of poverty. Furthermore, it is strongly implied that this cost-free path will be the result of scientists and engineers delivering miracle technologies that make emission-free energy, food, transport and materials cheaper than their dirty equivalents. Just throw money at the nerds and let them loose with some of the more presently unpalatable technologies – nukes, genetic engineering, carbon capture and storage. Do this and interminable arguments at climate forums are no longer necessary – our climate crisis will solve itself.

The appeal of this argument is obvious, and it seems to be popping up with increasing frequency. Charlton has clearly elucidated the difficulties of getting a binding deal on climate, particularly in the current circumstances. So throwing research-and-development money at our modern-day alchemists for the equivalent of a Philosopher's Stone sounds like a neat solution. However, as a long-time devotee of the black arts of engineering innovation, I fear it stands about as much chance of success in a relevant time-scale as did the alchemists themselves.

Helen Caldicott regularly mocks nuclear power with a quote attributed to Einstein – that it is a "hell of a way to boil water." It is cheap mockery, but there is a serious point to be made. With the exception of solar cells and tidal power, every source of energy of practical importance is tapped by means of a heat engine[1] – a device for turning temperature differences into mechanical energy. While virtually all small-scale heat engines are internal combustion engines, at the large scale two designs predominate: the steam turbine and the gas turbine. Whether the heat comes from burning wood shavings, the earth's geothermal heat, coal, running oil-filled pipes past curved mirrors in a desert, or splitting atoms, the process of turning that heat into electricity is essentially the same.

In the case of coal, the vast majority of the cost is not in combusting the coal itself, but in the heat engine that turns that combustion into useful energy. As such, in situations where coal is cheap and readily available, it is hard to imagine how replacing the coal burners with some other heat source could possibly make it cheaper to produce energy. Sure, an alternative heat source coupled with the most efficient heat engine currently available produces cheaper electricity than a coal-fired plant built today and using an old, inefficient heat engine design. But that does not help us find an alternative energy source that is cheaper than fossil fuel.

Similarly, while many of the green movement's objections to carbon capture and storage are overblown, there are fundamental reasons why CCS is unlikely *ever* to be price-competitive with conventional coal or gas without a carbon price or other subsidy measures. Again, proponents of the technology like to compare CCS plants using highly advanced heat engines that don't exist beyond the drawing board with the Beatles-era technology in many of the world's existing coal-fired power plants. What is usually omitted from such analyses is that these new heat engines can also be applied to coal combustion without the use of CCS, which is the relevant price comparison.

Charlton has already detailed the weaknesses of existing intermittent renewable energy technologies, and the even less developed technologies for large-scale energy storage that running an electricity grid based on intermittent renewables will require.

In the long term, there may be any number of technologies that square this circle, and our governments are still bizarrely reluctant to support their development in proportion to the need. But we simply cannot wait for the long term. The science of climate change makes amply clear that waiting another few decades guarantees catastrophe. Whether we like it or not, we will have to start acting seriously on climate change with the technologies we have now, or which are near deployment, and enhance them as we go.

So we're back to Charlton's dilemma, and our alchemists have told us to call them back in thirty years. I do not know what the way out is. But ultimately it is a political problem, and requires a political solution. Technologists are not going to bail us out of this one on their own, no matter how far we let them off the leash.

Robert Merkel

1 The energy that lifts water into clouds and propels the winds, tapped by hydroelectric and wind turbines respectively, is provided by heat from the sun.

Peter Karamoskos

Andrew Charlton makes a fundamental conceptual error in his essay that leads to several incorrect conclusions. The error is the conflation of electricity with energy and hence the incorrect assumption that most carbon emissions arise from electricity generation. Ipso facto, according to this line of argument, tackling climate change predominantly involves a race to replace coal with less carbon-intensive sources of electricity. If this were true, then meeting our emissions target would indeed be a herculean (and unlikely) task. But it is a flawed assumption, and as a consequence the conclusions are similarly compromised.

In 2009 (the most recent year for which the data is complete), 50 billion tonnes of carbon and its equivalents were emitted into the atmosphere. Of this, only 9 billion were due to electricity generation. Twenty-one billion tonnes were derived from transport and industrial sectors. A further 20 billion tonnes arose from deforestation and other gases (fugitive emissions in mining, agriculture and livestock, halocarbons, nitrous oxides and so on). Even in Australia, where coal accounts for 85 per cent of our electricity generation (compared with 41 per cent around the world), electricity is responsible for only 37 per cent of all greenhouse gas emissions.[1]

So what are the implications for how we tackle the challenge? Firstly, it is misleading to propose the electricity sector as the main avenue to carbon abatement. Charlton correctly points out how nearly impossible this would be, giving specific examples of the magnitude of the challenge that renewable electricity sources would have to meet, but this argument is based on a flawed premise. Even if we immediately eliminated all greenhouse gas emissions from electricity, we would be abating only 18 per cent of emissions worldwide on today's figures, well short of the 50 per cent required by 2050. Fossil fuels used in transport alone account for twice the emissions of electricity worldwide. Deforestation also probably accounts for twice the emissions of electricity.

These latter two categories are where most of the abatement can and must be achieved if we are to avoid a temperature rise greater than 2 degrees by the end of the century.

The false premise also undermines the nuclear industry's claim that nuclear power is essential to averting climate change because only it can hope to address carbon emissions from the electricity sector. Tony Benn, when he was Britain's energy secretary, used to warn against people who came to you with a problem in one hand and a solution in their back pocket. He learnt this from Britain's nuclear industry. The industry has never kept its promises, much less delivered on its claims. Whether it be electricity "too cheap to meter" in the 1950s or energy independence in the 1970s, it has never failed to disappoint. Climate change mitigation is the latest false promise. Charlton's expectation that the nuclear industry will double its capacity over the next few decades is pure fantasy. We currently commission about one new reactor a year somewhere in the world. If under the most optimistic conditions we raise that to eight a year for the next ten years and fifteen a year for the ten years after that, we have simply replaced the reactors that will be decommissioned by then. However, assuming that the nuclear industry pulled the proverbial rabbit out of a hat and was able to double its capacity over this time period, and (falsely) assuming that it generates no greenhouse gases itself, it would only abate an additional 2 billion tonnes of greenhouse gases per annum over the existing 2 billion it already abates, that is, a 4 per cent reduction of total global emissions in 2009.

Of course, we don't need arithmetic to arrive at this conclusion. The French nuclear power industry is a real-world example. France generates 80 per cent of its electricity from nuclear power. However, electricity accounts for only 21 per cent of all energy used. Fossil fuels account for 48 per cent, predominantly used in the transport and industrial sectors, with France consuming more oil per capita than the EU average, all of it imported.[2] So much for energy independence!

Technology and innovation have helped create the problem, so can they help solve it? Perhaps the better question is this: do we have anything else to rely on? The suggestion in the essay that there will always be some innovation around the corner that will adequately address today's problems misses the point. The application of such innovations, particularly when it comes to energy, is largely driven by politics. There might be a free market in ideas, but there is no economic free market when it comes to funding their application in such a strategically sensitive area as energy. The playing field is severely tilted in favour of fossil fuels, to the detriment of renewable technologies.

The fossil-fuel industry garners over US$400 billion per year in indirect government subsidies (through state spending) to cut the retail prices of gasoline, coal and natural gas around the world, according to the most recent figures from the International Energy Agency,[3] having risen 36 per cent in the past year alone. This compares to subsidies for the renewable sector (biofuels, wind power, solar energy) of US$66 billion, which have risen by only 10 per cent. In addition, the G-20 nations spent a further US$160 billion supporting the production and consumption of fossil fuels last year. The IEA goes on to state that these subsidies are "creating market distortions that encourage wasteful consumption … The costs of subsidies to fossil fuels generally outweigh the benefits." While governments argue that such subsidies are designed to help the poorest members of society, the IEA says that they generally fail to meet that goal.

Renewable technologies are playing catch-up, with a combination of belated and relatively small subsidies by comparison with those of the fossil-fuel and nuclear industries. Although existing technologies are further along the cost curve than renewable technologies, only the latter will provide sufficient cost reductions with scale to justify their subsidy. Fossil fuels, especially oil, are too subject to the pricing vagaries of the market, and economies of scale have long ago been achieved. I agree that the challenge is enormous, but the conclusions derived in the essay are inconsistent. Surely the technologies that continue to benefit enormously from economies of scale and technological innovation and are proven in the market (e.g. wind, solar, biomass), albeit still being more expensive than fossil fuels, have greater potential than "clean coal," which has never delivered, and nuclear power, which has never failed to disappoint? The market seems to think so.

The changes in renewable energy have been so rapid in recent years that perceptions can lag years behind reality. Renewables had grown to supply an estimated 20 per cent of global final energy consumption in 2010, growing strongly in all end-user segments: power, heat and transport. By the end of that year, renewables comprised one-quarter of global power capacity from all sources. Most technologies held their own, despite the global financial crisis, while solar PV surged, with more than twice the capacity installed than in the year before. No technology has benefited more than solar from the dramatic drop in costs. Despite the recession, total global investment in renewable energy broke a new record in 2010. Investment reached US$211 billion, up 32 per cent from the previous year. Renewable electricity generators provided in excess of half the world's new generation capacity (194 gigawatts) as of 2009–10. In 2010, renewables, excluding large hydro, surpassed nuclear's global capacity and received $151 billion dollars in private investment. Nuclear received no private investment. China's 2006 renewable electricity capacity (excluding large hydro) was seven times its nuclear capacity and was growing sevenfold faster; by 2010 that gap had widened, despite the world's most ambitious nuclear program. Renewables accounted for about 26 per cent of China's total installed electric capacity (nuclear only 1.1 per cent) and more than 9 per cent of total final energy consumption (not just electricity) in 2010.[4] Wind power alone in China is planned to account for a 100-gigawatt electricity-generating capacity by 2020, more than twice Australia's current capacity.

There may be many paths to a low-carbon future, but some, like nuclear power and carbon capture and storage, are no more than dead ends. We should save our scepticism for these rather than create false choices based on false premises.

Peter Karamoskos

1 Department of Climate Change, National Greenhouse Gas Inventory 2010, http://www. climatechange.gov.au/climate-change/emissions/progress.aspx.

2 Emission Database for Global Atmospheric Research (EDGAR) 2008, http://edgar.jrc. ec.europa.eu/overview.php?v=CO2.

3 http://www.worldenergyoutlook.org/.

4 Renewables 2011, Global Status Report, www.ren21.net/Portals/97/documents/GSR/ REN21_GSR2011.pdf.

Barney Foran

Andrew Charlton leaves the idealist in no doubt about the toughness and the luck required to propel planet-saving initiatives through the back rooms of power into legislation, and thence into our daily lives. The well-chosen vignettes on technological innovation urge us to remain optimistic about the man-made future he foresees. His research publications in global-scale economics are highly numerate and impeccable. However the biophysical scientist in me maintains that all economists must be made to jump through the hoops of physics and chemistry majors before they hang up their "practising economist" shingle.

The case is well made for the centrality of growth in energy supply to economic growth and thus development for the 6 billion in less-developed countries. High development measures are achieved at energy supplies of between 50 and 100 gigajoules (GJ) per capita and the rest is spent in shopping malls to maintain growth, not facilitate further development. The US and Australia consume about 300 GJ per capita and public policy wonders why lifestyle diseases are pushing the health system to the brink. Biophysical science knows that energy-driven growth cannot go on forever. Yet policy believes in a perfect way, a sweet spot, where we can all have our cake and eat it too. One physical solution is for the developed world to reverse its development pathway to the consumption levels of the 1980s, while developing nations reach their 80 GJ per capita as quickly and fluently as possible. Physical laws of thermodynamics form ultimate constraints to energy supply and use, and thus innovation must stall. Simply put, *First Law*: You can't win; *Second Law*: You can't break even; *Third Law*: You can't get out of the game.

Planetary scientists now see this man-made world (termed the "Anthropocene" geological era) is accelerating towards nine planetary thresholds, rather than the one relatively simple climate change threshold that befuddles world leaders. If Charlton thought Copenhagen was intractable, then to climate change he should add ocean acidification, ozone, atmospheric aerosols, freshwater use, land-use change, biodiversity loss, chemical pollution and the global phosphorus and nitrogen cycles. The nitrogen cycle boundary suggests another constraint for cornucopian policy, the Law of Unintended Consequences. In pursuing the belief that every problem has an eventual solution, Charlton charts the decline of mineral fertilisers essential in escaping Malthus's starvation trap, and the brilliant chemical synthesis by Fritz Haber and Carl Bosch of fixing atmospheric nitrogen to give the man-made urea and ammonium fertilisers that keep today's food production rampant.

This innovation, almost alone, stimulated global population growth to today's level since it allowed every woman on earth to pass the protein threshold necessary for successful breeding. (Sorry to all readers who thought that population growth was the Pope's fault.) More importantly, the Haber–Bosch process is so effective that the globe is now awash with nitrogen at more than double the background levels at the start of the twentieth century. Oceanic dead zones now abound off the shores of developed countries and are spreading to developing countries as relatively cheap fertiliser is used excessively and then transported by rivers to coastal zones. A biophysical reckoning says that the Haber–Bosch innovation mostly caused both the population explosion and the now rapidly expanding coastal dead zones. If this is progress, the analyst in me wonders how many big ones like this the globe can stand.

Most policy mainstreamers pat the heads of the peak-oil theorists and tell them "the stuff will never run out because modern economic functioning and growth would fail without it, and then where would the world be?" Serious analysts use metrics like the "energy profit ratio" or the "energy return on investment," which compare the energy invested in getting the stuff with the net energy out the other end of the oil refinery that ends up in the tank of the Commodore. In the heady days of the US east-coast oil boom, it was 1:100. That's one barrel of energy invested to 100 barrels returned. US and Australian oil now sit around 1:8 and both are on a steady decline waiting for the killer innovation that is thermodynamically limited, or the big field that sits under Antarctica. The physical point is that global resources are past the point of the original "cheap oil." This cheap oil catalyst enabled world development to accelerate in the 1950s after World War II gave us mass production of big diesel engines that cleared the bush, dug big holes and fished the world's oceans. Charlton is correct in saying we will not run out of the stuff! However, if we rely on the unconventional Canadian tar sands at an energy profit ratio of 1:3, the modern growth economy will soon choke on the physical inefficiency of this life cycle and its environmental impact.

Trade and globalisation are Charlton's forte, so it's a foolhardy scientist that takes him head-on. Adding a few nuances is necessary, though, particularly around the much-maligned Malthus. In addition to the fertiliser innovation that eased Europe's looming food constraints, surely it was colonisation and trade that added cheap staples of grain and meat to shopping baskets then, and so the system accelerates today. The physical reality of globalisation is that it allows affluent consumers to outsource their production impacts without conscience. Consumers in developed countries import ten food groups, causing large bio-diversity loss in developing countries, some obvious such as my morning coffee. The solution is not more of the same, but much less of it, perhaps with a cappuc-cino tax for every cup.

Charlton's effort at global synthesis is far-reaching, technically deep and, above all, honest. He really believes that complex solutions will evolve, and most importantly they must, for the good of humankind. Down in the laboratory where we face physical realities of thermodynamics and mass balance, we are not so sure. We know this era of economic growth and its sub-components have reached their zenith for the developed world. The next-world architecture is now required, and quickly.

Barney Foran

Andrew Charlton

Churchill famously described Britain's equivocal response to fascism in the 1930s as "the years the locusts hath eaten" – a lost opportunity to prepare for the great threat of that age. *Man-Made World* was motivated by a fear that we may look back on the first two decades of the twenty-first century with a similar sense of regret. The essay argues that we should – and can – do more to protect against the threat of climate change.

The nine published responses all broadly support this premise. Each correspondent urges us to slash our greenhouse gas emissions, but each has a different perspective on where the knife should fall.

Peter Hay believes we in the "affluent West" must bear the burden of climate change by scaling back our lavish lifestyles. The world cannot cut emissions, he maintains, unless rich countries lower their standard of living and accept "drastic cuts to [our] material opulence." For Hay, we Western consumers are prodigal sons: we have indulged in an orgy of materialistic excess, but now it's time to sober up and head back to the farm. He chides us for imperilling our planet through our intemperance, our "unconstrained market economy," "burgeoning cities" and the "capitalisation of agriculture." Like a character from Tolstoy, his heart yearns for a simpler time.

Hay's analysis is captive to sentiment, which is just as well, since his argument wouldn't survive outside the confines of his polemic. Tellingly, he can't, or at least doesn't, specify which elements of our current lifestyle we should part with.

Moreover, Hay gravely underestimates the value of economic progress. Progress has brought the health, education, wealth and opportunity that we dearly value in our daily lives. Progress has yielded the resources to defeat injustice and lift billions out of poverty.

The suggestion that we drastically cut our standard of living warrants Orwell's famous put-down: "One has to belong to the intelligentsia to believe things like that. No ordinary person could be such a fool." Indeed, "ordinary" voters are not fools. In country after country, voters have unequivocally vetoed any climate policy that has even a shade of a chance of seriously threatening the wellbeing of their families. Hay may want us all to live a simpler life, but abundant evidence suggests that most people in rich countries want continued material progress, and most in poor countries are desperate for it. The inescapable consequence is that we need to find workable solutions to climate change that do not severely compromise our prosperity.

John Quiggin also believes we should do more to address climate change but, unlike Hay, he doesn't believe that significant cuts to our living standards will be required. Quiggin is one of our finest and most practical economists. Australia is lucky to have a scholar of international standing who maintains a strong commitment to analysing domestic issues. Quiggin warns us not to be seduced by either the Cassandras or Pollyannas at the extremes of the climate debate. He says that tackling climate change will not be easy, but nor will it be impossible.

Quiggin believes we should be doing more to address climate change through market mechanisms. Specifically, he believes that a ramped-up emissions trading scheme with a carbon price of around $100 per tonne would go a considerable way towards decarbonising our economy. He shows, quite persuasively, that the decarbonisation challenge facing Australia is no larger than the economic transformation required to deal with our health and ageing challenges. Both will consume a similar proportion of our national income over the next half-century.

It is crucial to recognise that the total cost of decarbonising our economy will be a function of the available technology. If we had better technology – more efficient clean-energy alternatives, better batteries and electric cars, cheaper and safer nuclear power and effective technology to reduce the emissions from coal-fired power stations – the cost of decarbonising the economy would be relatively low. Without these technologies, the cost will be relatively high – indeed, I argued in the essay that without significantly better technology than we have today, it will be nearly impossible to decarbonise our energy, industrial and transport systems at a politically acceptable cost.

But how to promote technological advancement? Market mechanisms by themselves deliver too little technological progress because private companies prefer to invest in products and processes of innovation that they can quickly take to market. That is why most basic research is conducted in the government-funded university sector. For the same reasons, public funding is needed to support clean-energy research, which companies often consider too risky and too expensive. The essay argued that the best policies to stimulate new climate technologies therefore include a combination of price signals and direct government support.

Several respondents delve deeper into these technological issues. Robert Merkel believes that relying solely on improved technologies would be folly, because we don't know when investment in innovation will pay off and we cannot wait indefinitely. "Whether we like it or not, we will have to start acting seriously on climate change with the technologies we have now." Eric Knight takes the opposite view. He has a pure faith that technology will deliver us from climate change. To yoke ourselves to the engine of technological progress, all we need to do is embrace the market and its "clockwork of competition and reward." Ben McNeil is also a technophile. He believes that climate technology will not only cut emissions, but also deliver a new age of prosperity through a "clean industrial revolution." Climate change, he implies, is a welcome prod to lift the pace of human innovation. Like Mussolini's line on war, McNeil seems to believe that regular environmental disasters serve a useful historical purpose by keeping civilisation on its toes. While I'm not optimistic enough to believe that climate change will be a net economic positive in either the short or long term, I do sympathise with something in each of these three perspectives. Technology (as Knight says) holds huge promise, but (as per McNeil) it requires big investment and (as Merkel correctly points out) the research pay-off is uncertain. To manage these variables, we should begin reducing emissions now but simultaneously invest in new technologies for the future.

The essay also sought to emphasise the link between climate change and development – the twin challenges of our age. Across the table in Copenhagen we learned that the climate solutions proposed in rich countries – principally that we make fossil fuels more expensive and use less energy – do not work for poor countries, where more energy is needed and people don't have the capacity to bear greater costs. Since nine of the ten fastest emitters are poor countries, a truly global solution to climate change must include alternative strategies to reduce emissions in the developing world. Again, technological advances that make clean energy progressively cheaper will be an essential part of the solution for poor countries.

In this short reply I have tried to focus on the core themes of the essay and have fallen far short of addressing all the comments made in the responses, including strong points made by Peter Karamoskos, Barney Foran and Charles Berger. My thanks to all the respondents for their thoughtful and engaging words.

Andrew Charlton

Jeremy Baskin is a principal research fellow at La Trobe University and a former policy adviser to the post-apartheid government of South Africa.

Charles Berger is director of strategic ideas at the Australian Conservation Foundation.

Andrew Charlton was senior economic adviser to the prime minister from 2008 to 2010 and the prime minister's representative to the Copenhagen Climate Conference. He is the author of *Ozonomics* (2007) and *Fair Trade for All* (2005), co-written with Nobel laureate Joseph Stiglitz.

Barney Foran is a scientist who works at whole-economy analysis, blending financial and physical realities. After nearly thirty years with CSIRO, he is now a research fellow at Charles Sturt University in Albury.

Peter Hay is in the School of Geography and Environmental Studies at the University of Tasmania, and has advised the ALP at both state and federal levels. His books include *The Forests* (2007), with Matthew Newton, and *Main Currents in Western Environmental Thought* (2002).

Peter Karamoskos is a nuclear radiologist, a public representative of the Australian Radiation Protection and Nuclear Safety Agency and the Australian secretary of the International Campaign to Abolish Nuclear Weapons.

Eric Knight is a former Rhodes Scholar with a doctorate in economic geography, who has consulted for the OECD, the UN and the World Bank. He is the author of *Reframe* (2012) and writes for the *Sydney Morning Herald*, the *Age*, the *Drum*, the *Spectator* and the *Monthly*.

Anna Krien's first book, *Into the Woods* (2010), won the Queensland Premier's Literary Award for Non-Fiction and the Victorian Premier's People's Choice Award. Her writing has been published in the *Monthly*, the *Age*, *The Best Australian Essays*, *The Best Australian Stories* and the *Big Issue*.

Ben McNeil is a scientist and economist at the University of New South Wales and the author of *The Clean Industrial Revolution* (2009).

Robert Merkel is a lecturer in software engineering at Monash University. He writes for the Australian political blog *Larvatus Prodeo*.

John Quiggin is ARC Federation Fellow in Economics and Political Science at the University of Queensland. He has worked extensively on the economics of climate change and its implications for the Murray-Darling Basin. His most recent book is *Zombie Economics: How Dead Ideas Still Walk Among Us* (new edition 2012).

SUBSCRIBE to Quarterly Essay & SAVE nearly 40% off the cover price

Subscriptions: Receive a discount and never miss an issue. Mailed direct to your door.
- ☐ **1 year subscription** (4 issues): $49 a year within Australia incl. GST. Outside Australia $79.
- ☐ **2 year subscription** (8 issues): $95 a year within Australia incl. GST. Outside Australia $155.
- * All prices include postage and handling.

Back Issues: (Prices include postage and handling.)

- ☐ **QE 1** ($10.95) Robert Manne *In Denial*
- ☐ **QE 2** ($10.95) John Birmingham *Appeasing Jakarta*
- ☐ **QE 4** ($10.95) Don Watson *Rabbit Syndrome*
- ☐ **QE 5** ($12.95) Mungo MacCallum *Girt by Sea*
- ☐ **QE 6** ($12.95) John Button *Beyond Belief*
- ☐ **QE 7** ($12.95) John Martinkus *Paradise Betrayed*
- ☐ **QE 8** ($12.95) Amanda Lohrey *Groundswell*
- ☐ **QE 10** ($13.95) Gideon Haigh *Bad Company*
- ☐ **QE 11** ($13.95) Germaine Greer *Whitefella Jump Up*
- ☐ **QE 12** ($13.95) David Malouf *Made in England*
- ☐ **QE 13** ($13.95) Robert Manne with David Corlett *Sending Them Home*
- ☐ **QE 14** ($14.95) Paul McGeough *Mission Impossible*
- ☐ **QE 15** ($14.95) Margaret Simons *Latham's World*
- ☐ **QE 16** ($14.95) Raimond Gaita *Breach of Trust*
- ☐ **QE 17** ($14.95) John Hirst *"Kangaroo Court"*
- ☐ **QE 18** ($14.95) Gail Bell *The Worried Well*
- ☐ **QE 19** ($15.95) Judith Brett *Relaxed & Comfortable*
- ☐ **QE 20** ($15.95) John Birmingham *A Time for War*
- ☐ **QE 21** ($15.95) Clive Hamilton *What's Left?*
- ☐ **QE 22** ($15.95) Amanda Lohrey *Voting for Jesus*
- ☐ **QE 23** ($15.95) Inga Clendinnen *The History Question*
- ☐ **QE 24** ($15.95) Robyn Davidson *No Fixed Address*
- ☐ **QE 25** ($15.95) Peter Hartcher *Bipolar Nation*
- ☐ **QE 26** ($15.95) David Marr *His Master's Voice*
- ☐ **QE 27** ($15.95) Ian Lowe *Reaction Time*
- ☐ **QE 28** ($15.95) Judith Brett *Exit Right*
- ☐ **QE 29** ($16.95) Anne Manne *Love & Money*
- ☐ **QE 30** ($16.95) Paul Toohey *Last Drinks*
- ☐ **QE 31** ($16.95) Tim Flannery *Now or Never*
- ☐ **QE 32** ($16.95) Kate Jennings *American Revolution*
- ☐ **QE 33** ($17.95) Guy Pearse *Quarry Vision*
- ☐ **QE 34** ($17.95) Annabel Crabb *Stop at Nothing*
- ☐ **QE 36** ($17.95) Mungo MacCallum *Australian Story*
- ☐ **QE 37** ($20.95) Waleed Aly *What's Right?*
- ☐ **QE 38** ($20.95) David Marr *Power Trip*
- ☐ **QE 39** ($20.95) Hugh White *Power Shift*
- ☐ **QE 41** ($20.95) David Malouf *The Happy Life*
- ☐ **QE 42** ($20.95) Judith Brett *Fair Share*
- ☐ **QE 43** ($20.95) Robert Manne *Bad News*
- ☐ **QE 44** ($20.95) Andrew Charlton *Man-Made World*

Payment Details: I enclose a cheque/money order made out to Schwartz Media Pty Ltd. Please debit my credit card (Mastercard or Visa accepted).

Card No. ☐☐☐☐ ☐☐☐☐ ☐☐☐☐ ☐☐☐☐

Expiry date ___ / ___   Amount $ _____

Cardholder's name _____   Signature _____

Name _____

Address _____

Email _____   Phone _____

Post or fax this form to: Quarterly Essay, Reply Paid 79448, Collingwood VIC 3066 / Tel: (03) 9486 0288 / Fax: (03) 9486 0244 / Email: subscribe@blackincbooks.com Subscribe online at **www.quarterlyessay.com**